ROCKVILLE CAMPUS LIBRARY

D1711761

Bertin · Graphics and Graphic Information-Processing

Jacques Bertin

# Graphics and Graphic Information-Processing

Translated by
William J. Berg and Paul Scott

Walter de Gruyter
Berlin · New York 1981

*Title of the Original French Edition*

La Graphique et le Traitement Graphique de l'Information

avec la collaboration de

Serge Bonin, Jean-Daniel Gronoff, Alexandra Laclau,
Aline Jelinski, Madeleine Bonin, Georgette Couty, Paulette Dufrène,
Nancy François, Elise Rayez et tous les membres du
Laboratoire de Graphique

Copyright © by Flammarion, Paris, 1977

*Author*

Jacques Bertin
Directeur d'Etudes à l'Ecole des Hautes Etudes en Sciences Sociales
Directeur du Laboratoire de Graphique

*Translators*

William J. Berg
Professor at the Universitiy of Wisconsin

Paul Scott
Professeur au Conservatoire des Arts et Metiers

*Technical Assistence*

Howard Wainer, Ph.D.
Senior Research Associate at The Bureau of Social Science Research, Inc.
under a grant from The National Science Foundations (soc 76-17768)

John Wander, Ph.D.
Copy Editor, de Gruyter Publishers, Inc.

*Library of Congress Cataloging in Publication Data*

> Bertin, Jacques, 1918-
>   Graphics and graphic information-processing.
> 
>   Translation of La graphique et le traitement
> graphique de l'information.
>     1. Graphic methods.  2. Graphic methods--Data
> processing.  i. Title.
> QA90.B4713    001.4'226    81-12610
>   ISBN 3-11-008868-1     AACR2

Copyright © 1981 by Walter de Gruyter & Co. Berlin 30.
Publisher's Impressum see page 274.

# TABLE OF CONTENTS

### A. POSTMORTEM OF AN EXAMPLE
1. The stages of decision-making.......................... 2
2. The aim of graphics: a higher level of information
   - 2.1 Useful information............................... 11
   - 2.2 Information levels............................... 12
   - 2.3 Measurement of useless constructions............. 15
3. The three successive forms of graphic application
   - 3.1 The matrix analysis of a problem................. 17
   - 3.2 Graphic information-processing................... 20
   - 3.3 Graphic communication........................... 22
   - 3.4 Outline of work................................. 22

### B. GRAPHIC CONSTRUCTIONS
1. A "synoptic" of graphic constructions................. 24
2. Permutation matrices
   - 2.1 The reorderable matrix........................... 32
   - 2.2 The weighted matrix............................. 60
   - 2.3 The image-file.................................. 70
   - 2.4 The matrix-file................................. 86
   - 2.5 The array of curves............................. 90
3. Ordered tables
   - 3.1 Tables with 1, 2 or 3 characteristics............ 100
   - 3.2 Superimpositions and collections of tables....... 123
4. Reorderable networks................................. 129
5. Ordered networks: topography and cartography
   - 5.1 Information provided by a map................... 139
   - 5.2 The base map.................................... 141
   - 5.3 Cartography with one ordered characteristic..... 145
   - 5.4 Cartography with several characteristics........ 152

### C. THE GRAPHIC SIGN SYSTEM
(A Semiological Approach to Graphics)
1. Specificity of graphics............................... 176
2. The bases of graphics................................. 180
3. Variables of the image: the plane, size and value..... 186
4. Differential variables................................ 213
5. The law of visibility................................. 228
6. Summary............................................... 230

### D. THE MATRIX ANALYSIS OF A PROBLEM AND THE CONCEPTION OF A DATA TABLE............ 233
1. The apportionment table............................... 235
2. The homogeneity schema................................ 240
3. The pertinency table.................................. 245
4. Applications of matrix analysis....................... 251

CONCLUSION................................................ 265

| J | F | M | A | M | J | J | A | S | O | N | D | | |
|---|---|---|---|---|---|---|---|---|---|---|---|---|---|
| 26 | 21 | 26 | 28 | 20 | 20 | 20 | 20 | 20 | 40 | 15 | 40 | 1 | % CLIENTELE FEMALE |
| 69 | 70 | 77 | 71 | 37 | 36 | 39 | 39 | 55 | 60 | 68 | 72 | 2 | % —″— LOCAL |
| 7 | 6 | 3 | 6 | 23 | 14 | 19 | 14 | 9 | 6 | 8 | 8 | 3 | % —″— U.S.A. |
| 0 | 0 | 0 | 0 | 8 | 6 | 6 | 4 | 2 | 12 | 0 | 0 | 4 | % —″— SOUTH AMERICA |
| 20 | 15 | 14 | 15 | 23 | 27 | 22 | 30 | 27 | 19 | 19 | 17 | 5 | % —″— EUROPE |
| 1 | 0 | 0 | 8 | 6 | 4 | 6 | 4 | 2 | 1 | 0 | 1 | 6 | % —″— M.EAST, AFRICA |
| 3 | 10 | 6 | 0 | 3 | 13 | 8 | 9 | 5 | 2 | 5 | 2 | 7 | % —″— ASIA |
| 78 | 80 | 85 | 86 | 85 | 87 | 70 | 76 | 87 | 85 | 87 | 80 | 8 | % BUSINESSMEN |
| 22 | 20 | 15 | 14 | 15 | 13 | 30 | 24 | 13 | 15 | 13 | 20 | 9 | % TOURISTS |
| 70 | 70 | 75 | 74 | 69 | 68 | 74 | 75 | 68 | 68 | 64 | 75 | 10 | % DIRECT RESERVATIONS |
| 20 | 18 | 19 | 17 | 27 | 27 | 19 | 19 | 26 | 27 | 21 | 15 | 11 | % AGENCY —″— |
| 10 | 12 | 6 | 9 | 4 | 5 | 7 | 6 | 6 | 5 | 15 | 10 | 12 | % AIR CREWS |
| 2 | 2 | 4 | 2 | 2 | 1 | 1 | 2 | 2 | 4 | 2 | 5 | 13 | % CLIENTS UNDER 20 YEARS |
| 25 | 27 | 37 | 35 | 25 | 25 | 27 | 28 | 24 | 30 | 24 | 30 | 14 | % —″— 20-35 —″— |
| 48 | 49 | 42 | 48 | 54 | 55 | 53 | 57 | 55 | 46 | 55 | 43 | 15 | % —″— 35-55 —″— |
| 25 | 22 | 17 | 15 | 19 | 19 | 19 | 19 | 20 | 19 | 22 | | 16 | % —″— MORE THAN 55 —″— |
| 163 | 167 | 166 | 174 | 152 | 155 | 145 | 170 | 157 | 174 | 165 | 156 | 17 | PRICE OF ROOMS |
| 1.65 | 1.71 | 1.65 | 1.91 | 1.90 | 2. | 1.54 | 1.60 | 1.73 | 1.82 | 1.66 | 1.44 | 18 | LENGTH OF STAY |
| 67 | 82 | 70 | 83 | 74 | 77 | 56 | 62 | 90 | 92 | 78 | 55 | 19 | % OCCUPANCY |
| | | x | x | x | | | | x | x | x | x | 20 | CONVENTIONS |

**1**

⬇

**2**

| | | |
|---|---|---|
| J FM AM J J A S O N D  J FM AM J J A S O N D | 19 % OCCUPANCY<br>18 LENGTH OF STAY | ACTIVE<br>AND SLOW PERIODS |
| | 20 CONVENTIONS<br>8 BUSINESSMEN<br>11 AGENCY RESERVATION<br>4 SOUTH AMERICA | DISCOVERY FACTORS |
| | 12 AIR CREWS<br>13 CLIENTS UNDER 20 YEARS<br>16 CLIENTS MORE THAN 55 YEARS<br>14 CLIENTS FROM 20-35 YEARS<br>1 FEMALE CLIENTELE<br>2 LOCAL CLIENTELE | RECOVERY FACTORS<br>WINTER |
| | 7 ASIA<br>9 TOURISTS<br>10 DIRECT RESERVATION<br>17 PRICE OF ROOMS | WINTER-SUMMER |
| | 6 MIDDLE EAST, AFRICA<br>3 U.S.A.<br>5 EUROPE<br>15 CLIENTS FROM 35-55 YEARS | SUMMER |

Once upon a time there was a manager of a large hotel, anxious to improve his establishment's performance. He had his staff compile various statistics. The table of figures **(1)** remained on his desk for a number of days.

Then one day his assistant presented him with a graphic **(2)** constructed from the data table. After a few moments' attention, the manager summoned his staff, and with them
— defined a new price structure,
— modified the services offered to the guests,
— reorganized supplies,
— and modified his promotion campaign.
Then he rounded off the day with a visit to the mayor of the city, whose duties included the scheduling of conventions.

The results of his efforts assured him of rapid promotion.

This example demonstrates that it is not sufficient to have data, to have statistics, in order to arrive at a decision. Items of data do not supply the information necessary for decision-making. What must be seen are the relationships which emerge from consideration of the entire set of data. *In decision-making the useful information is drawn from the overall relationships of the entire set.*

This example also shows that graphics can uncover these overall relationships. That is its purpose. In contrast with pictography, graphics is not an art. It is a strict and simple system of signs, which anyone can learn to use and which leads to better understanding. Thus, it leads to better decision-making.

# A. POSTMORTEM OF AN EXAMPLE

*1. THE STAGES OF DECISION-MAKING*

Let us take a closer look at the hotel manager's problem.
It enables us to schematize the stages of analysis and decision-making.
It illustrates the role of graphics in these stages.
It gives a visible form to the problems encountered in information-processing.
It indicates the means of defining a useful graphic construction.

## 1.1   1st Stage: Defining the Problem

The manager hopes to improve the operation of his hotel. What decision must be made? To decide is to choose, and to choose he must first be "informed." He clarifies his problem by asking certain questions: Is full occupancy guaranteed? Are there slow periods? When? Where does the clientele come from in summer? In winter? Who are they? The problem is defined by progressively simpler questions which permit the composition of a list of potentially useful items of information. Note that this list of basic questions and useful information is purely a problem of imagination which no machine could solve. This first and fundamental stage in decision-making cannot be automated.

## 1.2   2nd Stage: Defining the Data Table

This stage is limited by the means and time available. Will the entire list of imagined questions be retained? Will the information be gathered by month, by week, or by day, each multiplying the work involved? A list of twenty monthly indicators seems sufficient here.
Once the useful statistics are defined in nature and number, the assistant gathers the necessary figures from the data in the hotel records.
Will he construct several tables, one for where the guests come from? another for their age? a third for data related to operating the hotel, etc. . . . ? No. He decides on a single table. Let us imagine his reasoning:

# The stages of decision-making

| J | F | M | A | M | J | J | A | S | O | N | D | | |
|---|---|---|---|---|---|---|---|---|---|---|---|---|---|
| 26 | 21 | 26 | 28 | 20 | 20 | 20 | 20 | 40 | 15 | 40 | | 1 | % CLIENTELE FEMALE |
| 69 | 70 | 77 | 71 | 37 | 36 | 39 | 55 | 60 | 68 | 72 | | 2 | % —"— LOCAL |
| 7 | 6 | 3 | 6 | 23 | 14 | 19 | 14 | 9 | 6 | 8 | 8 | 3 | % —"— U.S.A. |
| 0 | 0 | 0 | 0 | 8 | 6 | 6 | 4 | 2 | 12 | 0 | 0 | 4 | % —"— SOUTH AMERICA |
| 20 | 15 | 14 | 15 | 23 | 27 | 22 | 30 | 27 | 19 | 19 | 17 | 5 | % —"— EUROPE |
| 1 | 0 | 0 | 8 | 6 | 4 | 6 | 4 | 2 | 1 | 0 | 1 | 6 | % —"— M.EAST, AFRICA |
| 3 | 10 | 6 | 0 | 3 | 13 | 8 | 9 | 5 | 2 | 5 | 2 | 7 | % —"— ASIA |
| 78 | 80 | 85 | 86 | 85 | 87 | 70 | 76 | 87 | 85 | 87 | 80 | 8 | % BUSINESSMEN |
| 22 | 20 | 15 | 14 | 15 | 13 | 30 | 24 | 13 | 15 | 13 | 20 | 9 | % TOURISTS |
| 70 | 70 | 75 | 74 | 69 | 68 | 74 | 75 | 68 | 68 | 64 | 75 | 10 | % DIRECT RESERVATIONS |
| 20 | 18 | 19 | 17 | 27 | 27 | 19 | 19 | 26 | 27 | 21 | 15 | 11 | % AGENCY —"— |
| 10 | 12 | 6 | 9 | 4 | 5 | 7 | 6 | 6 | 5 | 15 | 10 | 12 | % AIR CREWS |
| 2 | 2 | 4 | 2 | 2 | 1 | 1 | 2 | 4 | 2 | 5 | | 13 | % CLIENTS UNDER 20 YEARS |
| 25 | 27 | 37 | 35 | 25 | 25 | 27 | 28 | 24 | 30 | 24 | 30 | 14 | % —"— 20-35 —"— |
| 48 | 49 | 42 | 48 | 54 | 55 | 53 | 57 | 55 | 46 | 55 | 43 | 15 | % —"— 35-55 —"— |
| 25 | 22 | 17 | 15 | 19 | 19 | 19 | 19 | 20 | 19 | 22 | | 16 | % —"— MORE THAN 55 —"— |
| 163 | 167 | 166 | 174 | 152 | 155 | 145 | 170 | 157 | 174 | 165 | 156 | 17 | PRICE OF ROOMS |
| 1.65 | 1.71 | 1.65 | 1.91 | 1.90 | 2 | 1.54 | 1.60 | 1.73 | 1.82 | 1.66 | 1.44 | 18 | LENGTH OF STAY |
| 67 | 82 | 70 | 83 | 74 | 77 | 56 | 62 | 90 | 92 | 78 | 55 | 19 | % OCCUPANCY |
| | | x | x | x | | | | x | x | x | x | 20 | CONVENTIONS |

What do all my data have in common? The monthly figure. So I can put the months along the abcissa (on *x*), place an indicator on each row of the ordinate (on *y*), and record the figures in the boxes or "cells."

Let us note the unity of the data table; indeed, the use of a single table indicates that the problem is well defined. Suppose that the assistant had also included the number of rooms, beds, people, repairs per floor. If this were added to the preceding information, the entire set of data could not be recorded in the form of a single table. If we attempted to use this set as a basis for analysis, we would be mixing two completely unrelated problems. Something still encountered all too frequently. This single table proves the homogeneity of the problem. It is the basis for all analysis, conscious or not. It is the means of organizing research. It is the starting point for any information-processing. The dimensions of the table and the nature (ordered, reorderable, topographic) of its entries determine the processing method and graphic. A single table may be impossible to construct for practical reasons: how could the fifty million inhabitants of a country be recorded in a single row? However, nothing prevents us from imagining such a construction and therefore "seeing" a given problem in the form of a table. This is the *matrix analysis of a problem*. It enables us to organize work.

Let us note that the information which preceded decision-making can always be *written or imagined in the form of a single table of numbers* (cardinal, ordinal or binary numbers without any labels). This principle will be a unifying concept throughout the book.

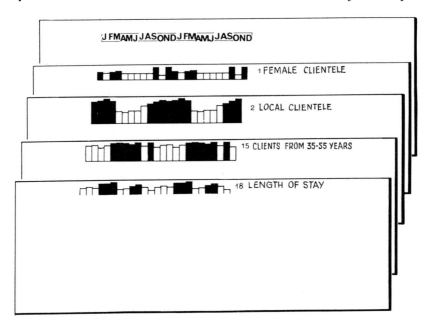

*The stages of decision-making* 5

## 1.3  3rd Stage: Adopting a Processing Language

The data table remained on the manager's desk for a number of days. Undoubtedly, because it was difficult to read and did not display the information, that is, the relationships useful for analysis and decision-making.
Indeed, the reading of the table of numbers provides no more than a linear sequence of details. The relationships, the similarities among these details, can only arise in memory, with no certainty that all the similarities have been perceived. Furthermore, they appear less clearly as the data increase, and generally the amount of data considered serves as a measure of a project's credibility.

What then is the assistant to do? Sparing his own efforts while remaining concerned with efficacy, by *graphic transcription* he ensures the possibility of making all the similarities appear. However, he takes two essential precautions:

*He ensures maximum visual efficacy.* He transforms each row of numbers into a profile. He shades in the months where the number exceeds the annual average in order to display the variations clearly. He records the twelve months twice so that a potential cycle does not risk being split up and thus overlooked.

*He ensures the mobility of the image.* He constructs each profile on a separate card (image-file). Thus he draws each profile only once, but he remains free to class the profiles in different ways and to construct different images from the same drawing.

This is a fundamental point, because it is the internal mobility of the image which characterizes modern graphics. A graphic is no longer "drawn" once and for all; it is "constructed" and reconstructed (manipulated) until all the relationships which lie within it have been perceived. The practical possibilities for permuting the elements on a diagram are numerous. The means employed by the assistant are simple and within anyone's reach. For the permutation of rows and columns, special equipment is available. Remember that we live in an age of computers and electronic display screens, and that all such permutations can now be carried out by pressing a button.

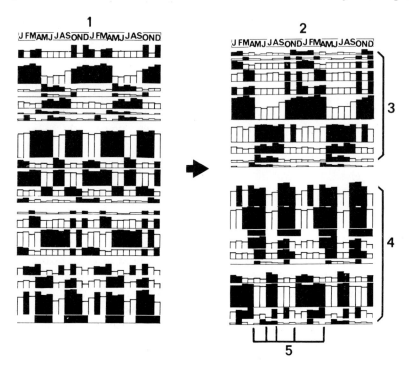

## 1.4 4th Stage: Processing the Data: Simplifying Without Destroying

When the cards are classed in the same order as the original table, they construct image **(1)**. If it stopped there, the graphic transcription would be absolutely useless. However, since the rows are physically independent, they may be reclassed to make the similarities appear. It is sufficient to study the cards two by two and to group those which resemble one another **(1→ 2)**.

The assistant then discovers: a) that the indicators construct two systems of variation, one semi-annual **(3)** the other quarterly **(4)**, and b) that the year can be broken down into four highly differentiated periods **(5)**. Image **(2)** is the *simplification* of image **(1)**.

# The stages of decision-making

At this point the reader will no doubt ask, "How, without any calculation or predefined system, can image **(2)** be derived from image **(1)**?" This question reveals the originality of modern graphics. In fact, classing things is a common and continual operation. To understand and act is to categorize and class. But according to what system? Acquired habits, alphabetical order, educational classifications, different systems of measurement? In any case, according to a linear system which precedes the drawing, since to draw is to fix an order once and for all. In this sense it is inconceivable to reclass what is already drawn.

*However, everything changes if the drawing is physically reclassable.* Indeed, visual perception is spatial perception and it allows anyone to use a new system of classing: the simultaneous consideration of several different elements. Since this exercise is hardly ever suggested to us, not only are we not practiced at it, but we are even unaware of its existence. However, it is a completely natural exercise, and a nine-year-old child could easily construct image **(2)**. In only a few hours, under proper conditions, an adult can relearn to "see" and rediscover that the eye is made to perceive similarities and sets, not just signs, words, and numbers.

And what does this prove? We can state that the simplification is no more than regrouping similar things. The eye simplifies by correcting the irregularities it notices in the initial disorder. Indeed, the original inventory is a disorder, produced by the random nature of human imagination and the contingencies of general classifications. The eye simplifies. This means that it eliminates differences of position, "visual distances" which signify nothing. The permutation of lines removes everything which hides the specific inherent organization created by the finite set of data. When this organization emerges, it permits subsequent discussion, not about the organization (which only depends on the data), but:
1. about the nature of the data considered at the outset;
2. about the modifications which would ensure a better understanding of the discovered information.

Let us observe, however, that at the moment of simplification there is no need to refer to the written notations in the legend, that is, to the nature of the concepts, in order to discover the specific order. It depends only on the figures, on the profiles' shape. This operation may therefore be entrusted to a machine; this stage can be automated.

*Postmortem of an example*

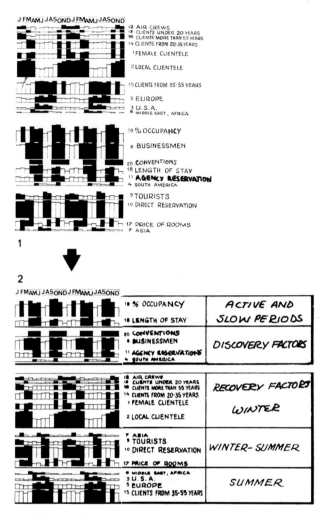

## 1.5 5th Stage: Interpreting and Deciding . . . . or Communicating

From the moment that the similarities become apparent and create the overall complex of information, the assistant may choose the indicators *which interest him* from the perspective of decisions to make. It is only now that he must be able to read the meaning of each row and each column in order to define his problem. Which is? To ensure maximum occupancy. To eliminate slow periods.

## The stages of decision-making

To pose the problem, *he modifies the simplified image* **(1→2)**. He brings those indicators which indicate general activity to the top: percentages of occupancy, length of stay. They indicate the extent to which the hotel is filled and, at the same time, the slow periods, which fluctuate quarterly.
Three other indicators have the same profile. They tend therefore to create or to sustain this fluctuation. These are the "discovery factors." The indicators that have an inverse or different profile are obviously those which could be used to work against slow periods. These are the "recovery factors." They form three groups, characterized by their frequency.

The assistant and afterwards the manager modified the simplified image. This means that they proceeded to certain choices, for which they took into account:
1) *Intrinsic information,* that is, the internal relationships revealed by the image;
2) *Extrinsic information*, that is, the nature of the problem and the interplay of the intrinsic information with everything else. And, by definition, everything else is that which *cannot be processed by machine.* Extrinsic relationships cannot, by definition, be automated. They are, however, of fundamental importance in interpretation and decision-making. Thus, the most important stages—choice of questions and data, interpretation and decision-making—can never be automated. There is no artificial "intelligence."

With this image before him the manager can make decisions. He discovers the characteristics of slow periods: absence of conventions, reduction in the percentage of businessmen, and in the role of travel agencies. But he also discovers that there are marked differences between the summer and winter guests during these slow periods. In winter: more local guests, a higher percentage of women, greater differences in age. In summer: more foreign guests and a more homogeneous age group. These observations add together clearly in the diagram and construct an image of the guests for each period of the year. It is then relatively easy to reorient the promotion campaign and to better structure the services and supplies. Careful analysis reveals that the return sought through variable pricing could be better adjusted. But is this not a discovery factor? Careful analysis also reveals that the December convention does not have the same influence as the others and that moving it or changing it could be beneficial.

*Postmortem of an example*

**1**

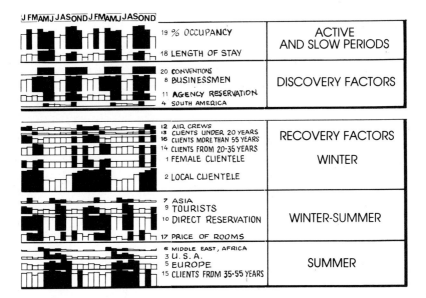

**2**

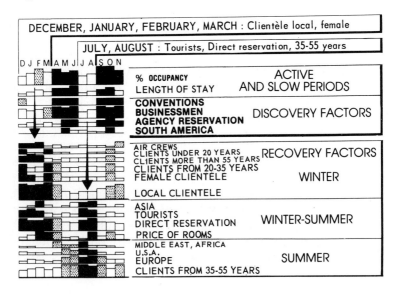

*The stages of decision-making* 11

With this image before him the manager can also decide . . . . not to make any decisions and to complete this information by investigating new indicators or by refining the data, down to the weekly level for example. In this case, he would return to the second stage and pick up the cycle from that point, normal procedure in thorough research.

With this image before him, finally, the manager can decide to inform the staff, to convey to them the results of this study. But in that case the image would be too complicated. He simplifies it in order to underline what is important to see, and he works out an explanatory text **(2)**. *A graphic designed for communication highlights the main points* of the results of the investigation.


## 2. THE AIM OF GRAPHICS: A HIGHER LEVEL OF INFORMATION

### 2.1 Useful Information

*Information is the reply to a question*

"In May, which type guest is most common?"
"Guests over 55, when are they most common?"
In any data table there are *two types of question*: questions *introduced by* x, that is by the objects; questions *introduced by* y, that is by the characteristics.
The result is obvious: any graphic construction which does not enable us to find a visual reply to the two types of question provides only one part of the information. To be useful, a construction must above all supply a visual answer to both types of question. *This principle immediately eliminates all graphic constructions which destroy any entry categories used in constructing the data table.*

*Useful information involves regrouping*

One may wonder why the table of figures remained on the manager's desk for a number of days, while the simplified image **(1)**, although less precise, enabled decisions to be made. It was the discovery of the periods and their characteristics that made it possible to make decisions, as if reading 12 x 20, 240, precise details was less useful than the recognition of four periods and five groups of characteristics **(1)**.

To have useful information meant reducing twelve months to four periods and twenty characteristics to five groups of characteristics. *Useful information involves a pertinent reduction in the length of the entries in the data table.* This observation is the key to any statistical operation.

Useful information does not correspond to the general categories: age, profession, hometown, etc., which enabled us to conceive the data table but on the contrary, to the *new groupings defined by the set of relationships constructed by the interplay of the data.*
"How do the months and characteristics group themselves in relation to the entire set of data?" This is the most useful question, the one which enables us to make decisions.
For the notion of quantities of information, we must therefore substitute the notion of levels of information, approached verbally by levels of questions and graphically by levels of reading.

## 2.2 Information Levels

Information is a relationship. But this relationship can exist among elements, subsets or sets. And these three levels must be retained in the subsequent graphic.

*The elementary level*

"In May, there are 15 guests older than 55." This information is the relationship which exists between one element of $x$ and one element of $y$. We cannot descend to a more elementary level in the data table. There are as many items of elementary information as there are entries in the table. This level is the saving grace of bad constructions, almost always affording a visual reply.

*The intermediate level*

"What happens in winter?" This information is the relationship which exists between the subset "December, January, February" and the corresponding groupings of characteristics. There is as much intermediate information as there are imaginable subsets. The subsets can be defined in two ways:
- a priori and verbally: what happens in winter?
- a posteriori and visually: what are the characteristics of the semi-annual

*The aim: a higher level of information* 13

**THREE LEVELS OF INFORMATION**

elementary question — intermediate question — overall question

TWO TYPES OF QUESTIONS

in X

in Y

system *constructed by the data*? These observations show that graphic processing can work with prior hypotheses but that it can also function independently of them.

*The overall information level*

"What different periods are generated by the entire set of characteristics?" The relationship between the set of the "time" component and the "characteristics" component: the highest level of knowledge attainable from the data table. As we have seen, it is this overall level which is necessary for decision-making. To reach it is the main purpose of graphics. We can state that when the construction involves this level of reading, the lower levels are also legible, whereas the opposite is not true.

Consequently, a construction which does not enable us to define groupings in $x$ and $y$ does not reach the overall information level of the entire set. This is the mark of its inefficacy.

# Postmortem of an example

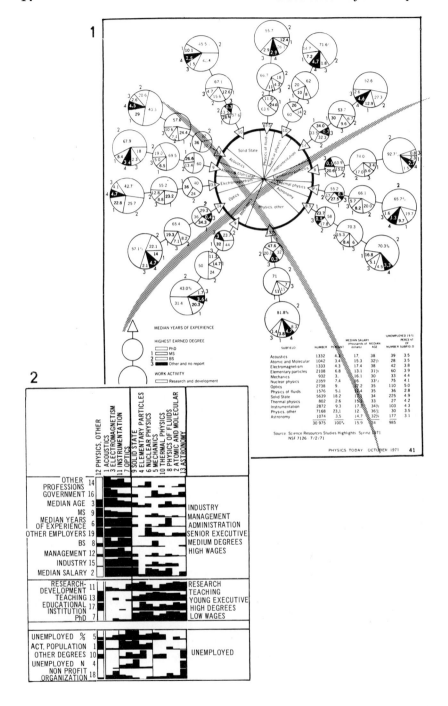

*The aim: a higher level of information* 15

## 2.3 Measurement of Useless Constructions

*Measurement*

The notion of information levels enables us to accurately measure the efficacy of the many imaginable graphic formulae. How long does it take to define the groups formed by $x$ and $y$? In numerous cases, several hours are required, and indeed any reply becomes impossible when it is necessary to memorize a large number of elementary data items. A "standard" construction, on the other hand, provides an almost instantaneous answer.

*The efficacy of a graphic construction is revealed by the level of question which receives an immediate response.* Outside of standard constructions, most constructions provide a visual reply only at the level of elementary information.

*The standard construction*

This is simply a matrix construction, that is, an $xy$ construction which conforms to the data table, but with:
1st — numbers transcribed in $z$ by a visual variation ordered from white to black or from small to large;
2nd — rows and/or columns reclassed in such a way that the groupings are revealed.
How can we even come to terms with a set of data, much less discover the groupings which the data construct in $x$ and $y$ — the aim of the entire operation — when the designer separates the two entry categories from the table? Consider the following example.

*A useless construction*

Diagram (1) illustrates an article concerning employment trends in physics in the U.S.A. What is it about? Can the reader ask even a single pertinent question? This diagram remains on the elementary level. Nevertheless, the information which it contains is not uninteresting. A standard construction (2) shows that in the U.S.A. industrial employment, administration, higher management, lower degrees . . . . and higher salaries go together, while the lower salaries are associated with research, teaching and higher degrees. However, it also shows that unemployment follows quite another pattern. This is what is revealed by the overall level of information, where all the categories can be compared. Once this level is reached, any information at the elementary level becomes more interesting, either as evidence of the general tendency or as an exception (see page 40).

*The graphician's responsibility*

Any problem can be conceived in the form of one double-entry table and thus can be transcribed in the form of a matrix. This is the most common solution, one which makes full use of the properties of visual perception. Essentially, it varies only in relation to the dimensions of the data table. Graphics is the visual means of resolving logical problems. An overly complex construction is a poorly defined equation, and as such a mistake. Thus, contrary to what is often thought, graphics is a very simple and efficacious sign system which anyone can put to use.

A graphic is not only a drawing; it is a responsibility, sometimes a weighty one, in decision-making. A graphic is not "drawn" once and for all; it is "constructed" and reconstructed until it reveals all the relationships constituted by the interplay of the data. The best graphic operations are those carried out by the decision-maker himself.

A graphic is never an end in itself; it is a moment in the process of decision-making. To construct a useful graphic, we must know what has come before and what is going to follow. The example of the hotel enables us to identify the *successive forms of graphic application*. Any problem can be related to this general schema.

## 3. THE THREE SUCCESSIVE FORMS OF GRAPHIC APPLICATION

The successive stages in decision-making correspond to very different forms of graphic application.

| STAGES IN DECISION-MAKING | GRAPHIC APPLICATION |
|---|---|
| - Define the problem<br>- construct the data table | Matrix analysis of a problem (define the questions) |
| - Adopt a processing language<br>- process the data (classify the *comprehensive* data) | Graphic information-processing (discover the answers) |
| - Interpret<br>- decide or<br>- communicate the *simplified* data | Graphic communication (communicate the answers, if need be) |

*The three forms of graphic application*  17

## 3.1  The Matrix Analysis of a Problem and the Construction of a Data Table

In defining a problem one can only have recourse to imagination. A series of questions, as simple as possible, will permit formulating of hypotheses and foreseeing which data will have to be compiled. This first stage is no doubt the most difficult, since imagination cannot be learned. It illustrates the difference between man and machine and often indicates the true competence of an investigator.

The dimensions of a study vary greatly. With some 240 elementary items of data, the hotel manager can make pertinent decisions. In such cases, the form of the data table is almost self-evident. Nevertheless, this table results from a series of calculations and choices which will serve later to show how the matrix analysis of the problem can be useful at this stage. In a much larger study—one without direct precedent and involving numerous and varied data—the construction of the table depends not only on the data but especially on the hypotheses and the available means for reducing these data. These means are the mathematical and graphical methods of information-processing. However, in considering hypotheses and methods, it is necessary to envisage the whole problem. The matrix analysis of a problem is a process which enables us to see the whole, that is, to construct it graphically, and to "foresee" the possible choices and their repercussions. This process is further developed on p. 233; the essential points are given here.

*The homogeneity of a study*

A study is only homogeneous if all the elements have a point of comparison. This point of comparison has a graphic form: it is the component which is common to all the other components in the study. It is that component which, if placed on $x$, becomes the STATISTICAL OBJECT for which all the other components are CHARACTERISTICS which can be placed on $y$.

To plan a study is first to conceive this single $xy$ table. To plan a study is, for example, to see that a series of tables which compare components (C)

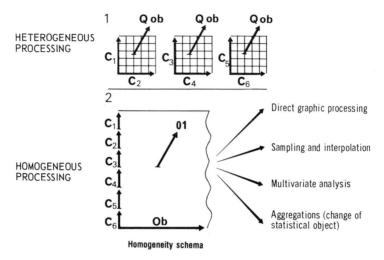

two by two **(1)** is a heterogeneous study, even though the common point in all the tables is the same "population." Indeed we must not confuse the "point in common" with the "point of comparison." The comparison of all these tables could only be verbal.

For the study to be homogeneous, for the comparison of the tables to be based on a logical demonstration, it is necessary to construct a point of comparison. It is necessary to construct schema **(2)**, where the "population of objects" (Ob) provides the point of comparison. This is a homogeneity schema, whose purpose is to schematize the data matrix.

*Necessary choices*

The homogeneity schema **(2)** sets out the elements of the problem.
1st) It enables us to see *if the data are homogeneous,* that is, if a point of comparison really exists among all the data. We often encounter studies in which the point of comparison does not exist or is not constructed **(1)**.
2nd) It enables us to calculate the magnitude of the problem and to compare it with the time and means available. How many investigators spend their time accumulating data without considering the time and means necessary to process them. They end up by processing only a minute portion of the comparisons at their disposal, and since they have not per-

*The three forms of graphic application*                                    **19**

ceived the entire set of possible comparisons, their choice is not supported by a logical demonstration.

3rd) It enables us to *choose a processing method* **(2)**. The magnitude of the problem can be compatible with direct processing, with *graphic processing* for example. But when it exceeds the limits of direct processing, a choice must be made:

- Are we going to keep the same point of comparison and proceed by *sampling* and interpolation? But on what basis do we carry out the sampling?
- Are we going to keep the point of comparison and, with the aid of a computer, use *multivariate algorithms*? But which ones? What loss of information would be entailed?
- Are we going to *aggregate* the data? Then we must *choose a new point of comparison*, a new "statistical object." Which one?

4th) It enables us to *record the hypotheses* in a way which is appropriate to the data. How can we entertain all the equivalent choices? By recording the hypotheses in such a way that one can see if the choice of processing method will exclude the answer to a given question. A "pertinency table" completes the homogeneity schema and enables us to display the hypotheses.

*The method of matrix analysis*

This method calls for the drafting of three documents:

a) The apportionment table inventories all the components which we intend to consider.

b) The homogeneity schema schematizes the data matrix constructed by these components and enables us to define the processing stages.

c) The pertinency table checks the relationship between the retained data and the hypotheses and determines the calculations which must precede the final drafting of the data table.

The drafting of these documents is further developed on p. 233.

*TERMINOLOGY*

A series of aggregations transforms the data. We will see on p. 254 how a component such as "different professions" can pass through various statistical situations.

A COMPONENT, that is, a set, a differential concept:
- can be a CHARACTERISTIC attributed to certain objects,
- and become a point of comparison, that is, a STATISTICAL OBJECT to which characteristics are attributed,
- in order to ultimately become a QUANTITY OF OBJECTS which measures the relationship between two characteristics.

The apportionment table enables us to record these different situations.

When the OBJECTS are on $x$, each of the rows of the table is called a CHARACTERISTIC (age, income) or an INDICATOR (age group, income class).

## 3.2 Graphic Information-Processing

This is the second form of graphic application. The matrix analysis enabled us to define and construct the data table in relation to a mathematical or graphic method of data processing. The mathematical methods are well known and have been the subject of numerous publications. The same is not true for graphic processing. The hotel example shows that we can consider four stages.

*Choosing a graphic construction.* The graphic construction depends on two main factors: the number of characteristics, that is, the number of rows in the data table; and the reorderable, ordered or topographic nature of its components. The "Synoptic of Graphic Constructions" on p. 25 classifies constructions according to these two factors. It enables us to define the type of construction appropriate to each problem and will serve to organize our discussion in Chapter B.

*Transcribing the data.* The problems of transcription differ in difficulty according to the particular case and are considered separately for each construction.

*Simplifying the data.* The aim of simplification is to make the relationships appear, that is, to *display* sought after information. In problems involving more than three components, simplification is achieved by the transformation of the image. Visual permutations are based on the eye's ability to perceive spatial entities. The eye perceives sets and can thus compare and bring together two similar rows by disregarding meaningless distances. Modern equipment, often very simple, makes these permutations possible for nearly anyone.

*Interpreting . . . and deciding.* Interpretation is a modification of the simplified image. It takes into account:
1st - The *intrinsic* information, that is the groupings discovered through simplification. This simplification can be automated.
2nd - The *extrinsic* information, that is the nature of the problem and the interplay of the information with everything else. Now, this "everything else" is that which cannot be automated. Interpretation reinserts the finite set just processed into the indefinite set from which it had been extracted. This indefinite set only makes sense in the imagination of the "interpreter" and within the scope of his knowledge.

It is here that the originality and specificity of graphics appears: the seeing from the whole to the detail and from the detail to the whole. There is no better method of stimulating the imagination, of discovering the most pertinent questions, of defining new processing methods, of organizing the final presentation. The mathematical manipulation itself is enhanced by the image. *The image is the core of the "interactive" or transactional phase*, that is, of the phase involving the choice of new operations. This finding is recent and makes us realize that the computer is not conceived for the image. A factor analysis, for example, must be redrawn by hand to be usable. It is impossible to display a matrix of average dimensions in the conditions required for its use. Trichromatic superimposition is unthinkable. The $z$ dimension of the image, which must not be confused with perspective or color, remains, in fact, unexploited. Most computer systems can only display lines.

To coordinate, at the time for decision, the properties of the image with those of the computer, the system should be reconsidered from scratch, by redistributing the properties of visual perception, by introducing the third or "z" dimension of the image, and by facilitating permutations. This amounts to the conception of a real "interactive or transactional graphic computer."

Graphics offers the means of going beyond what can be automated. We can find numerous methods for automating the diagonalization of a matrix, but we will never find methods for automating the conception of the most useful subsets. Graphic permutations can only be carried to their conclusions by the "decision-maker" himself.

Interpretation is fulfilled by decision-making, which involves several options:
- action, that is, building, buying, decreeing;
- resuming the study with new data, which leads back to the second stage;
- communicating the discovered relationships to others.

For this communication one can—but not have to—employ graphics, in which case the graphic problem becomes quite different.

## 3.3 Graphic Communication

When the hotel assistant drew up his image-file he did not know the relationships that the data would reveal. He used graphics to discover these relationships, in order *to discover what should be said* or done. This is graphic processing. When the teacher draws a diagram or a map on the blackboard, he knows in advance the relationships he is going to construct. He uses graphics *to tell others* what he has already discovered. This is graphic communication.

*Graphic processing* involves two imperatives which do not apply to graphic communication:
- it must transcribe all the data from the table, that is, the *"comprehensive"* data;
- it must answer all the pertinent questions and allow the two components of the data table to be simplified.
Graphic processing poses problems of dimensions and manipulations.

*Graphic communication* involves transcribing and telling others what you have discovered. Its aim: rapid perception and, potentially, memorization of the overall information. *Its imperative: simplicity.* This simplicity of forms authorizes the superimposition of images. Graphic communication poses problems on the level of simplification and selectivity.

We should avoid confusing graphic processing and graphic communication, drawing simplified maps when comprehensive studies are necessary, or superimposing comprehensive documents in a vain attempt to increase the informational value. Illegibility, rapidly reached, renders the graphic useless.
It is generally easy to pass from a comprehensive image to graphic communication, to construct an "interpretation matrix" from a comprehensive matrix (pp. 57, 89, 167). The opposite is obviously impossible.

## 3.4 Outline of Work

The logical sequence of graphic operations is presented on the opposite page. However, to apply it, to choose between mathematic and graphic processing, for example, it is necessary to become acquainted with the principal graphic constructions — their potentials as well as their limitations. This is the aim of chapter B. In chapter C, approaching graphics as a sign system enables us to summarize the main observations made in A and B. Chapter D is devoted to the matrix analysis of a problem.

## The three forms of graphic application                                      23

**MATRIX ANALYSIS**

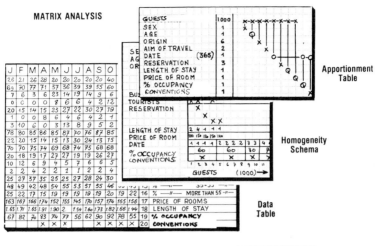

Apportionment Table

Homogeneity Schema

Data Table

**GRAPHIC INFORMATION-PROCESSING**

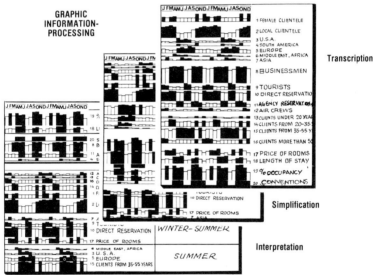

Transcription

Simplification

Interpretation

**GRAPHIC COMMUNICATION**

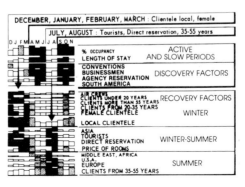

# B. GRAPHIC CONSTRUCTIONS

## B.1. A "SYNOPTIC" OF GRAPHIC CONSTRUCTIONS

### B1.1 An impassable barrier

With up to three rows, a data table can be constructed directly as a single image, producing a *scatter plot* or correlation diagram, in which the objects are in $z$. However, an image has only three dimensions. And this barrier is impassable. Consequently, with more than three rows, there are only two ways of denoting comprehensive information:
- constructing several scatter plots and sacrificing the overall relationships of the entire set;
- placing the objects on $x$ and the characteristics on $y$, that is, *constructing a matrix*. The overall relationships are then discovered through permutations.

*Let us consider, for example, a table with two rows* **(4)**: towns A, B, C, etc., for which we know the price of bread and the cost of housing. Two constructions are possible:
- placing the objects A, B, C on $x$ and the characteristics on $y$ **(5)**, a *matrix* construction. It involves classing A, B, C according to a characteristic ( ∽ );
- placing the objects in $z$. They become points, which enables us to place the two characteristics along $x$ and $y$, respectively **(6)**. This is a *scatter plot*. It classes the objects directly and makes apparent both their groupings and the relationship between the two characteristics.

The same applies for a table with three rows **(7)**. In scatter plot **(9)** the third row is expressed by the size of the points.
These two types of construction distinguish a "repartition" **(2)** from a "distribution" **(3)** for tables having one row.

# A "synoptic" of graphic constructions

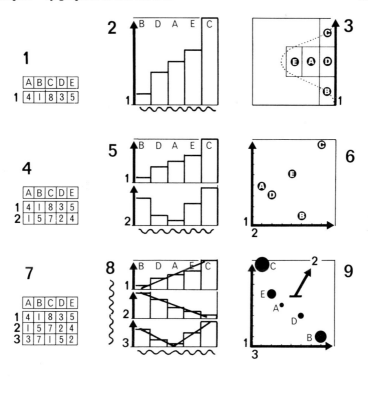

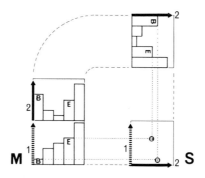

*Relationship between the matrix construction and the scatter plot.* This figure shows how columns B and E in the matrix construction (M) become points B and E in the scatter plot (S).

*Yet, a table with more than three rows* **(10)** cannot be constructed *directly* set up as a single signifying image. What we have to construct will be a series of scatter plots two by two **(12)**, or a series of scatter plots with three characteristics **(13)\***, or a superimposition of scatter plots, which means a superimposition of images **(14)**. In all these cases, the overall relationships are lost. The image has only three dimensions.

There remains only the matrix construction **(11)**, which enables us to discover the overall relationships *through permutation* (∼∼∼). And this is dynamic graphics, the extension and fulfillment of static graphics.

### B1.2 The unity of a problem.

The impossibility of *directly* constructing a table having more than three characteristics in a single image often leads to sub-dividing a given problem. Take the study of prices in thirty-one major cities (p. 43). This study has been commented on throughout the world, but the commentaries are all anecdotal: two towns or two products compared in relation to the seventeen published tables. However, the information itself is a unified whole, constructing only one table, one matrix that reveals the overall relationships, such as the socio-economic systems and ways of life that give each anecdote its full meaning.

And this is the real problem: is the anecdote significant or is it an exception? And if an exception, an exception in relation to what? In relation to general tendencies, to overall groupings? It is these that our memory searches for. It is these that we must SEE.

*Thus the main problem we have to work at is a problem with* n *characteristics. The principal construction is the matrix construction. Scatter plots are merely exceptional constructions; above all,* a problem with *n* characteristics is not the sum of *n* problems with two characteristics.

This is the reason that the classification of the *synoptic* on p. 29 is based on problems with *n* rows, since problems with one, two or three rows are only special cases of a problem with *n* rows. The habitual going from

---

\* For a table with *n* rows there are $\frac{n(n-1)}{2}$ scatter plots with two characteristics and $n - 2$ comparable scatter plots with three characteristics, two of which are held constant.

# A "synoptic" of graphic constructions

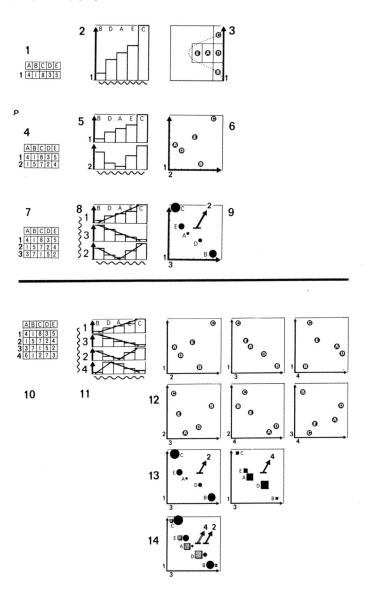

what seems the simplest (a table with one row) to what seems the most complicated (a table with *n* rows) only results in the subdivision of research and decision-making, and destroys our vision of the overall problem: a complaint which can well be lodged against classical graphics.

## B.1.3 The synoptic

The synoptic on the facing page displays all the basic graphic constructions as they relate to the data table.
Take a data table containing objects A, B, C . . . along $x$ and characteristics 1, 2, 3 . . . of these objects along $y$.
From this data table it is easy to observe:
1) the number of rows, that is, the number of characteristics;
2) the nature of the series of "objects," which, for example, is ordered, (0), for months; reorderable, ($\neq$), for individuals; or topographic, (T), for townships. The synoptic classifies graphic constructions according to these two principles.

*Tables with more than three rows*

A table with more than three rows (n) leads to permutable matrix constructions (∼∼∼), in which the characteristics are always reorderable ($\neq$).

*If the objects are $\neq$, it is a table ( $\neq \neq$ ). The construction is the reorderable matrix,* **(1)**, permutable in $x$ and $y$. The "weighted matrix" and the "matrix-file" are special cases of it (pp. 61 and 86).

*If the objects are ordered,* **(0)**, it is a table ( $\neq 0$) which has two basic constructions: the image-file **(2)** and the array of curves **(3)**, applicable when the slopes are meaningful.

*The ordered table* **(9)**, like the map **(18)**, constructs a fixed-reference image. *A collection of tables* **(4)** or a *collection of maps* **(5)** enables us to class the images according to a given similarity as well as to define groups of characteristics and objects. Remember that the superimposition of several tables or maps in a single figure leads to elementary reading and does not answer the fundamental question: what groups are formed by the $x$'s and/or the $y$'s of the data table?

*Tables with one to three rows*

Tables with one to three rows offer two basic constructions: *the matrix construction* places A, B, C on $x$ and leads to matrices with three, two or one row **(6, 7, 8)**, all of which entail a reclassing of the objects. Constructions **(13, 14, 15)** are arrays of curves; *the scatter plot* places A, B, C in $z$; when the data are quantitative, it makes a *direct construction*, without reclassing possible, **(9, 10, 11, 12)**.

In topographies, bi- or tri-chromatic superimposition **(16, 17)** reveals the overall relationships.

*Networks (N) and topographies (T)*

A network portrays the relationships which exist among the elements of a

# A "synoptic" of graphic constructions

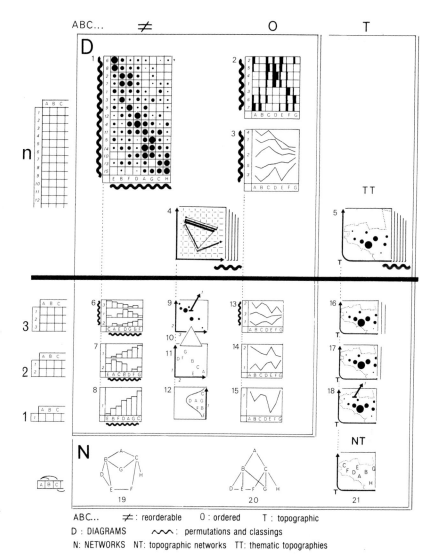

ABC... ≠ : reorderable    O : ordered    T : topographic
D : DIAGRAMS    ∿∿ : permutations and classings
N: NETWORKS    NT: topographic networks    TT: thematic topographies

single component. This component can be ≠, and the network is transformable on the plane **(19)**. It can be 0, and the network is transformable on one dimension of the plane **(20)**. Finally, it can be a topography **(21)**, that is, a non-transformable network: an ordered network.

But any network can be constructed in a matrix form. The elements are transcribed twice: once along *x*, and once along *y*. The relationships become points, and the matrix is permutable.

# Graphic constructions

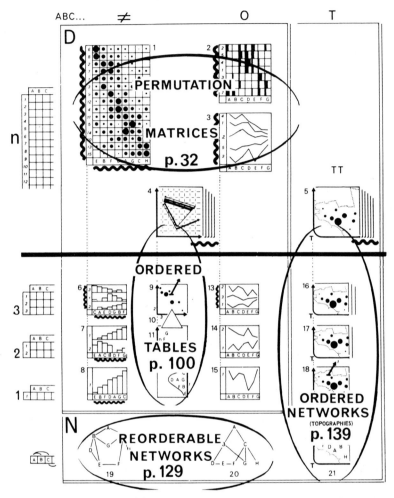

ABC...≠: reorderable    O: ordered    T: topographic
D: diagrams    ∼∼∼: permutations and classings
N: networks    NT: topographic networks    TT: thematic topographies

The table above refers to the pages in this book which discuss the various constructions.

## B.1.4. Utilization of the synoptic

First let us remember that the cells of the data table must only include yes/no answers, ordinal numbers, or quantities. However, the scatter plots corresponding to one, two and three rows imply that these rows do not have yes/no answers.

All the tables can contain symbols indicating the absence of data— which is different from zero—or "inapplicable" data.

# A "synoptic" of graphic constructions

*To refer to the synoptic* we start with the number of rows and the nature of the components in the data table. The synoptic then indicates the basic construction corresponding to the structure of the data.

*The choice of a different construction* must be justified. The constructions given on the synoptic are those capable of reaching the overall level of information. To adopt a different construction amounts to a shift in relation to the synoptic and a reduction in the perceptible information level. A problem involving $n$ rows does not correspond to $n$ problems involving one row. Consequently, a shift must be justified by the analysis of which questions will not be answered rather than by acquired habits. In any case the synoptic provides a reference point for discussing a given shift.

*The choice between a map and a diagram* rests on the balance between the length of the topographic component A, B, C and the number $n$ of characteristics. A large number $n$ leads to a matrix, a more powerful, supple and precise instrument for processing than the map, which can only be introduced at a later stage. To manifest relationships of proximity, the groupings discovered through processing are projected onto the map (pp. 50, 83, 138, 167).

On the other hand, if A, B, C is extensive, even infinite (as with a topographically continuous phenomenon), a series of maps combined with tri-chromatic superimposition is justifiable as a processing instrument.

*The limits of graphic information-processing.* Permutations are difficult when the following orders of magnitude are exceeded:

- reorderable matrix           $(\neq)$ 120 $\times$ $(\neq)$ 120
- (experimental equipment)     $(\neq)$ 500 $\times$ $(\neq)$ 100
- matrix-file                  $(\neq)$ 1000 $\times$ $(\neq)$ 30 non-permutable
- image-file, array of curves  $(\neq)$ 1000 $\times$ (0) unlimited
- collections of tables or maps $(\neq)$ unlimited

*Combinations of graphic processing*

In order to reduce information to the dimensions of a reorderable matrix, we must discover about 100 objects representative of all the objects. *The matrix-file* enables us to proceed with this selection (page 87) but with a maximum of about thirty characteristics. *The image-file* (page 70) and the *array of curves* (page 90) allow us to proceed with an ordered characteristic of any length whatever. Collections of tables and maps have no limits.

## B.2. PERMUTATION MATRICES

### B.2.1. THE REORDERABLE MATRIX

This is the basic construction. All the other standard constructions are only special cases of it. This construction is usable whenever the data table takes the form ( $\neq \neq$ ) and does not exceed approximately $x \times y = 10{,}000$. However, experimental processing has gone as far as $415 \times 76 = 31{,}850$ items of data.*

In this construction, all the rows are equal as are all the columns, which permits the use of permutation equipment. On the other hand, in the weighted matrix, the width of the rows and columns varies in relation to certain overall parameters.

#### B.2.1.1. Principle of matrix permutations

Take the following problem, intentionally a very simple one. For sixteen townships A, B, C . . . we know the presence or absence of nine characteristics 1, 2, 3 . . . . Should the same planning decisions be applied to all these townships?

The data construct table **(1)**. The two components of the table are reorderable. As in the hotel example, processing consists of bringing the similar rows together **(2→ 3)**, with the resulting much simpler image **(3)**. It remains nonetheless difficult to interpret, since the component A, B, C is not ordered. If we now adopt some technical means for bringing together the similar townships, that is, the similar columns, **(4→ 5)**, by cutting out the columns with a pair of scissors for example, image **(5)** is constructed quite easily. We thus discover the overall relationships and interpretation is simple **(6)**. In effect, within the limits of these data, we should apply the same decisions to townships H and K, based on the presence of characteristics 1, 3, 8 and on the absence of the others. However, these same decisions should not be applied to townships N, J, P, M, I, F, E, A, B. Finally, townships O, L, G, D, C justify a third sort of decision.

Visual reclassing has enabled us to define characteristic groups along with particular situations, such as township B, which can belong to two different groups. But do the township groupings revealed by the matrix also form regional geographic groupings? Here the matrix is mute. Only topography can supply the answer (see p. 139).

---

*Madeleine ROMER. "Application du traitement graphique de l'information en sociologie," *Revue française de sociologie*, XVI (1975), pp. 79-94.

# The reorderable matrix

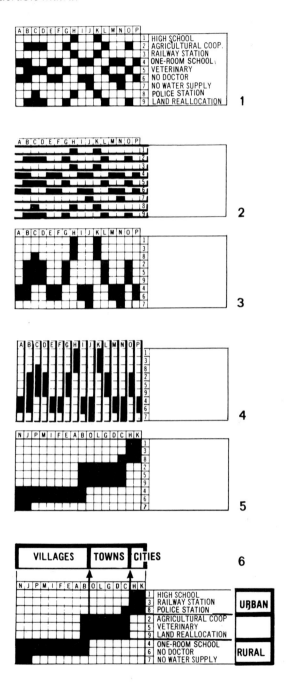

1

"Domino" No. 1

2

"Domino" No. 3 on a photocopier

3

Keyboard permutations on a cathode screen

*The reorderable matrix* 35

### B.2.1.2. Permutation equipment

We are concerned here with permuting rows, then columns, without having to re-draw the image each time. There are several possibilities.

*Cutting out and reclassing.* It is possible to permute a small-sized matrix without specialized equipment. Each element is cut out on relatively thick paper. It must be visibly identified by its row (in numbers) and its column (in letters). All types of permutation are possible. This method is especially appropriate for "weighted matrices" (p. 60).

*Specialized equipment.* Several permutation devices exist, but the only one which enables us to visualize the quantities is the "domino" apparatus perfected at the *Laboratoire de Graphique de l'Ecole des Hautes Etudes en Sciences Sociales*. There are three types. "Domino 1" does not exceed approximately 120 × 140 elements. "Domino 2" is heavy equipment, designed to experiment with very large matrices (up to 600 or 700 × 100). "Domino 3" is a miniaturized, portable apparatus.

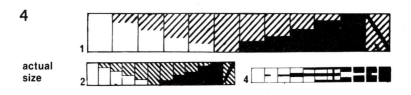

4

actual size

The "domino" device **(4)** includes eleven visual steps and special signs for noting missing or inapplicable data.* It is reversible: by turning over a row or column one displays its inverse. A set of rods threaded either through $x$ or $y$ makes permutations possible **(1)**. Photographic equipment should be used to record the principal stages of simplification and interpretation. "Domino 3" avoids this contingency, since it can be used with standard copying systems **(2)**.

*Cathode screen.* If one has access to a computer system with a display screen, the matrix can be shown on this screen and permuted at the control console **(3)**. However, this does not seem to be the best solution, at least with present systems, which cannot present the $z$ dimension of the

---

* Missing or inapplicable data cannot be represented as "black," nor therefore as "white." To be neutral, they must be gray and marked with a special sign.

image. Moreover, at the fundamental level of interpretation, one must have sufficient time to study the image and carry out the appropriate permutations.

### B.2.1.3. Preparatory tables

*The design of the data table*
The data table serves as a reference throughout the investigation and is generally published. It must therefore be designed with great care.
- Duplication by photocopying or other processes should be foreseen.
- Large-size paper should be used.
- Cells about 1 cm wide and 8 mm high should be traced in ink.
- On the table, as on the matrix, it is more important to read the definition of characteristics than of objects. Accordingly, the characteristics should be placed on *y* so that their definitions will be horizontal (see note on p. 251).
- Definitions are always in CAPITAL letters.
- Definitions and numbers must be unambiguous to an outside reader. They should be traced in pencil (corrections are always necessary) and "drawn" not written.
- The table can be set up in any order, but it is indispensable to number rows and columns.
- This table is going to be used a great deal, so copies should be made.

*The step table*
Since our permutation equipment involves a given number of steps, the data must be converted into step numbers (see p. 197).
To record these steps it is convenient to use a framework which conforms to the data table. Darken the lines every five or ten spaces. In each cell record very clearly in pencil the appropriate step number **(1)**. The matrix will be constructed from this table.

### B.2.1.4. Setting up the initial or "zero" matrix

The dominos are classed by step number. Each row is set up according to the step table, by placing the dominos on a board fitted with two stops **(2)** or by threading them directly onto a rod. When there are more than say 50 rows, it is advisable to place a reference domino every five or ten columns, corresponding to the darkened line on the table. Checking,

---

\* For choice and calculation of data, see chapter D.

## The reorderable matrix

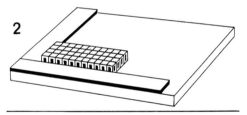

which is indispensable, will be much easier. These "reference columns" are removed afterwards. A special row and column of reversible white on black should be added to indicate potential inversions. The setting up of the matrix by two people is faster and more accurate.

*Titles and numbers* are drawn or typed on strips of paper which are then stuck onto ten or so dominos reserved for this purpose. Titles can be composed with ready-made letters or carefully written by hand. They must be in CAPITAL letters, and to be legible on photographic reductions, must be as large as possible. Avoid abbreviations unfamiliar to the layman.

When all the data are set up, the matrix reproduces the data table. This is the initial or "zero" matrix, which should be photographed for subsequent verification and evidence.

### B.2.1.5. Permutations

*Quantitative matrices.* Take the "zero" matrix on the facing page **(1)** and proceed in the following manner:

1st — Choose a row, if need be at random, but preferably one that includes few unknowns and few similar values, one that portrays a phenomenon whose general influence can be sensed. Take for example row four. Place it at the top. Thread the matrix vertically. Class row four. The entire matrix will then be classed along $x$ according to the ordered profile of row four **(2)**.

2nd — Thread the matrix horizontally. Bring together row four with those of similar profile, that is, one, twelve, eight. Place at the bottom those rows with a contrary profile: nine, two, eleven. Leave profiles unrelated to four in the middle, but group them according to their similarities: five/three and ten/seven. We thus obtain an overall diagonalization **(3)**.

3rd — Form systems, that is, associate opposite groups **(4)**:
system I: type four and its opposite type nine,
system II: type three and its opposite type ten,
system III: the remaining profiles. In fact, they will form one or several new systems.

If we schematize the result, we see that in a very general way, the systems construct an $X$, then a $V$, then a $W$, etc. **(7)**. This is the basic form of matrix simplifications.

Figures **(5)** show the main *equivalences of form* which are the bases for permutations. Figures **(6)** involve the inverse (■) of certain parts, further simplifying the image. Figures **(8)** remind us that the nearly uniform rows or columns (white or black) must be repositioned near the edges, so as to exclude them from the overall perception of differentiated rows. They may be returned to their places after processing.

*Binary (yes/no) matrices.* The processing of binary matrices proceeds by partial regrouping. Interest is focused on a precise mark that the permutations progressively augment. Other marks, black or white, are created and augmented. It is with these marks that we later construct a diagonal or systems.
In a matrix which involves both quantities and binary numbers it is advisable to group the quantities and to process their set first.

# The reorderable matrix

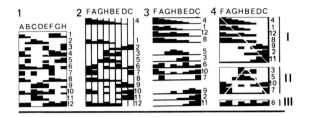

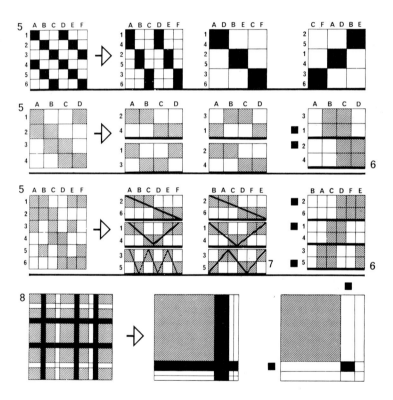

*Exclusions.* Permutations create groupings. However, certain rows or columns do not fall into these groupings and must be set aside **(8)**. They either represent special cases in relation to the considered set (p. 59) or a new system which will be simplified later (pp. 40, 57).

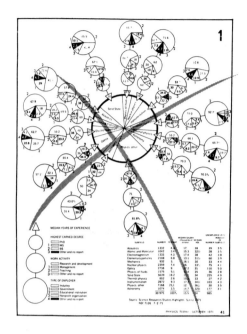

## B.2.1.6. EXAMPLES

*EMPLOYMENT TRENDS IN PHYSICS IN THE U.S.A. (1971).*
*MATRIX 19 × 13*

This is the example mentioned on page 15. To understand the meaning of diagram (1), the data table (2) must be reconstituted and set up as a reorderable matrix. The calculation of steps (p. 197) is done characteristic by characteristic constructing the "zero" matrix (3). Visual simplification leads to (4), which contrasts the bottom rows with the top rows. By bringing together these two contrasting groups, a "system" is constructed at the top of (5), and a second system emerges at the bottom of (5). Column twelve is a special case which does not fit into any of the groupings.
During these permutations, the meaning of rows and columns is intentionally ignored. The consequent groupings result *solely* from the numbers.

And what do these groupings mean? It is only now that we must look at definitions and begin "interpretation." First we notice that column twelve portrays the "other" items, which enables us to say that these are "truly other" and do not belong to either of the revealed systems.
The first system clearly divides the branches into two groups: acoustics, electromagnetics, instrumentation and optics are the appanages of older men, high salaries, management, and industry. The other branches are in the areas of research, teaching, advanced degrees and lower salaries. However, the second system, at the bottom, shows that one cannot rely upon this division in discussions of unemployment. It obeys other socio-economic laws whose salient features do not enter into this case and are undoubtedly unknown in classical statistics.

## The reorderable matrix

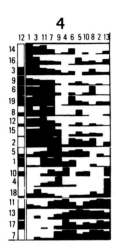

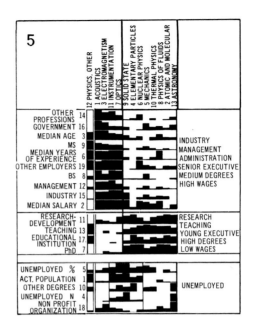

Diagram (1) shows that a good designer can be fully unaware of graphics and its use, while matrix (5) shows that it is not necessary to "know design" in order to make full use of the properties of graphics.

Prix des services                                                                                                           Tableau 14

| Services | Amsterdam | Athènes | Beyrouth | Bogotá | Bruxelles | Buenos Aires | Chicago | Copenhague | Düsseldorf | Genève | Helsinki | Hongkong | Johannesburg | Lisbonne | Londres | Luxembourg |
|---|---|---|---|---|---|---|---|---|---|---|---|---|---|---|---|---|
| | Fr. | Fr. | Fr. | Fr. | Fr. | Fr. | Fr. | Fr. | Fr. | Fr. | Fr. | Fr. | Fr. | Fr. | Fr. | Fr. |
| Coupe de cheveux (homme) | 3.50 | 3.47 | 6.65 | 1.88 | 4.77 | 4.30 | 12.95 | 8.60 | 4.89 | 5.— | 4.16 | 4.32 | 4.52 | 2.64 | 3.36 | 4.56 |
| Shampooing et mise en plis (dame) | 7.— | 8.67 | 6.65 | 5.88 | 11.— | 12.90 | 23.73 | 13.80 | 8.18 | 12.— | 9.85 | 8.64 | 9.— | 4.53 | 6.85 | 10.55 |
| Nettoyage chimique (complet 2 pièces) | 6.55 | 4.70 | 4.66 | 2.35 | 8.54 | 8.06 | 7.55 | 11.08 | 9.82 | 11.90 | 10.25 | 5.40 | 4.34 | 9.06 | 4.92 | 10.21 |
| Lavage et repassage d'une chemise | 1.78 | 1.58 | 1.— | –.94 | 1.30 | 1.61 | 1.30 | 1.59 | 1.58 | 1.80 | 1.90 | –.58 | 1.08 | 2.64 | 1.16 | 2.30 |
| Ticket de tram, bus, métro (simple course)[1] | –.60 | –.36 | –.33 | –.12 | –.70 | –.27 | 1.72 | –.72 | 1.— | –.50 | –.88 | –.58 | 1.08 | –.45 | 2.59 | –.43 |
| Raccordement téléphonique (taxe mensuelle) | 17.86 | 7.20 | 14.63 | 2.35 | 18.— | 10.70 | 24.16 | 20.14 | 21.38 | 8.50 | 4.65 | 14.40 | 9.— | 7.55 | 17.15 | 8.69 |
| Conversation locale (cabine publique) | –.12 | –.14 | –.33 | –.05 | –.43 | –.11 | –.43 | –.14 | –.24 | –.20 | –.52 | –.22 | –.30 | –.08 | –.26 | –.26 |
| Port (lettre, service intérieur) | –.30 | –.36 | –.33 | –.24 | –.30 | –.21 | –.26 | –.35 | –.36 | –.30 | –.52 | –.07 | –.15 | –.15 | –.22 | –.26 |
| Abonnement mensuel à un quotidien[2] | 7.14 | 10.12 | 9.98 | 7.05 | 7.80 | 8.49 | 9.70 | 16.37 | 8.— | 5.33 | 9.30 | 8.64 | 7.83 | 6.80 | 7.24 | 4.17 |
| Billet de cinéma[3] | 4.16 | 2.60 | 2.66 | 1.88 | 5.21 | 4.84 | 12.95 | 6.18 | 5.98 | 6.— | 4.40 | 2.88 | 6.62 | 3.78 | 7.11 | 3.91 |
| Total | 49.01 | 39.20 | 47.22 | 22.74 | 58.05 | 51.49 | 94.75 | 78.97 | 61.43 | 51.53 | 46.43 | 45.73 | 43.92 | 37.68 | 50.86 | 45.34 |

| Services | Madrid | Mexico | Milan | Montréal | New York | Oslo | Paris | Rio de Janeiro | Rome | São Paulo | Stockholm | Sydney | Tokyo | Vienne | Zurich |
|---|---|---|---|---|---|---|---|---|---|---|---|---|---|---|---|
| | Fr. | Fr. | Fr. | Fr. | Fr. | Fr. | Fr. | Fr. | Fr. | Fr. | Fr. | Fr. | Fr. | Fr. | Fr. |
| Coupe de cheveux (homme) | 3.50 | 6.90 | 5.69 | 8.30 | 12.95 | 8.90 | 3.90 | 5.70 | 4.80 | 5.69 | 9.35 | 7.24 | 6.05 | 5.— | 6.— |
| Shampooing et mise en plis (dame) | 3.72 | 10.35 | 9.09 | 16.60 | 25.89 | 11.77 | 11.73 | 8.55 | 7.90 | 9.49 | 11.22 | 16.89 | 7.26 | 7.50 | 12.— |
| Nettoyage chimique (complet 2 pièces) | 6.50 | 4.14 | 7.30 | 8.30 | 10.80 | 15.10 | 9.38 | 3.99 | 6.63 | 7.40 | 20.15 | 6.51 | 8.47 | 9.40 | 12.40 |
| Lavage et repassage d'une chemise | 1.08 | 1.38 | 1.61 | 1.45 | 1.95 | 2.54 | 1.92 | 1.42 | 1.83 | 1.99 | 2.93 | 1.69 | 1.21 | 1.36 | 1.60 |
| Ticket de tram, bus, métro (simple course)[1] | –.25 | –.17 | –.69 | 1.25 | 1.29 | 1.51 | –.55 | –.47 | –.69 | –.33 | 1.58 | –.97 | –.36 | –.83 | –.70 |
| Raccordement téléphonique (taxe mensuelle) | 7.95 | 13.45 | 8.53 | 27.28 | 30.20 | 18.11 | 15.63 | 13.30 | 10.06 | 33.22 | 7.76 | 34.50 | 13.31 | 16.67 | 8.35 |
| Conversation locale (cabine publique) | –.19 | –.07 | –.31 | –.41 | –.43 | –.19 | –.35 | –.05 | –.31 | –.47 | –.17 | –.24 | –.12 | –.17 | –.20 |
| Port (lettre, service intérieur) | –.12 | –.28 | –.34 | –.25 | –.26 | –.42 | –.31 | –.06 | –.34 | –.06 | –.46 | –.24 | –.18 | –.33 | –.30 |
| Abonnement mensuel à un quotidien[2] | 7.23 | 11.49 | 11.60 | 12.45 | 38.80 | 14.49 | 9.25 | 11.40 | 13.72 | 7.04 | 12.47 | 10.71 | 9.08 | 5.67 | 4.60 |
| Billet de cinéma[3] | 3.60 | 4.14 | 9.95 | 10.38 | 12.95 | 3.92 | 9.38 | 4.75 | 10.50 | 4.75 | 6.25 | 9.65 | 6.66 | 4.05 | 5.50 |
| Total | 34.14 | 52.37 | 55.11 | 86.67 | 135.52 | 76.95 | 62.40 | 49.69 | 56.78 | 70.44 | 72.34 | 88.64 | 52.70 | 50.98 | 51.65 |

[1] Valable pour l'ensemble du réseau urbain  [2] Taxe de factage comprise  [3] Catégorie de prix moyenne, pour la séance du soir dans un cinéma présentant des films en exclusivité
Etat: juillet 1970

## PRICES IN THIRTY-ONE MAJOR WORLD CITIES IN 1970.
## MATRIX 129 × 31

*The information* above enables us to compare thirty-one world cities in terms of prices, income, taxes, holidays and working hours. It was published by the Union des Banques Suisses in the form of seventeen annotated data tables. The problem of comparing them found no better solution than indexing prices "per 100 at Zurich." Indeed this problem was not the major concern of the editor, who sometimes placed cities along *x*, sometimes along *y*, and even divided certain tables into two parts (1)! Accordingly, it is not a surprise that the international press, while recognizing the value of this information, only published simplistic and anecdotal commentaries!

*Further reckoning* increases our interest in the document considerably. The prices are in Swiss Francs and thus of interest only to the international traveler. However, for each city the document provides the incomes of school teachers, bus drivers, mechanics, bank clerks, and secretaries. These data enable us to calculate the cost of living, not in Swiss Francs, but

## The reorderable matrix

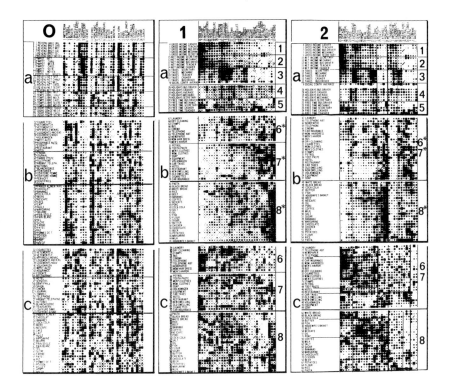

in relation to the average income of these five professions. A Volkswagen costs 8,414 Swiss Francs in New York and 8,640 in Hong Kong, representing, for the average individual, 3.3 months of work in New York and 15.7 in Hong Kong! The prices, divided by the average net income in each city, indicate *the real cost of living in terms of man-hours* for workers in each country. As such they become interesting for hundreds of millions of individuals and enable us to explore various types of societies.

*A single table* places the thirty-one cities along $x$, while the seventeen original tables (a and c) and the real calculated costs (b) are brought together along $y$. Graphic transcription, in the order of the original document, constructs *matrix n° 0* (using "domino 3" equipment). How, faced with this confusion, may we avoid merely anecdotal commentary. *Matrix n° 2* is classed according to the prices in Swiss Francs (c) and will be of interest to international travelers (p. 47).

*Matrix n° 1* is classed according to the cost of food as a percentage of income **(8\*)** while retaining groups as defined by taxes **(3)**. The great differences in taxes (38% in Stockholm, 0% in Hong Kong) in fact represent a difference in the redistribution of income.* It

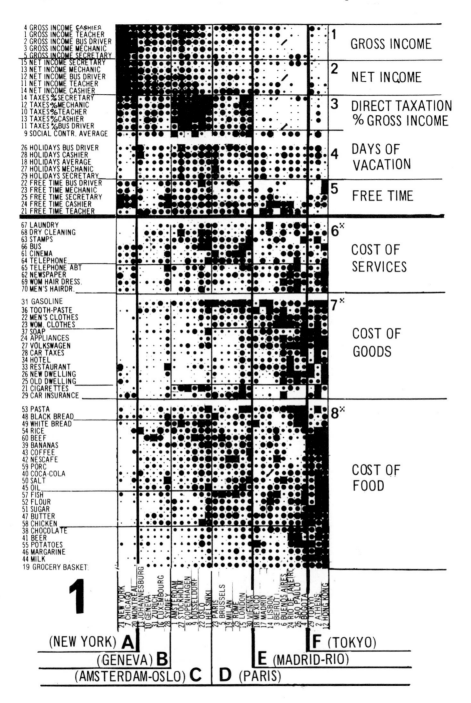

## The reorderable matrix

separates the taxed societies (A, B, C, D) from the others. In the former, costs range from low to average. They range from average to high in the latter. The data set constructs six basic types of societies.

- A. North American: highest income, lowest costs, very few vacations (4), but some free time (5).**
- B. Geneva, with Luxembourg and Johannesburg: almost similar to A, except for free time. Special case: Sydney.
- C. "Scandinavian" with Dusseldorf: maximum redistribution of income (3), long vacations and a lot of free time, low cost of goods (7*), higher cost of services (6*) and food (8*), particularly in Oslo and Helsinki.
- D. Relatively expensive European: special case: goods in London (7*).
- E. Hispano-American: separated from the others by low taxes and the high cost of goods, it declines from Mexico to Bogota.
- F. Tokyo with Athens and Hong Kong: the opposite of the New York type!

With these types, any anecdote becomes significant, whether it concern the general tendency set up or a special case: vacations in Paris, restaurants in New York and Sydney, cigarettes in Scandinavian countries, rice in London, etc. . . The graphic matrix enables anyone to undertake his own study and to evolve a "geography of consumption"! Table (9) shows real costs, calculated in percentage of average net monthly income and classed according to matrix n° 1.

* Row (9): "social assessments" is included in (3). It simply indicates the administrative means of collection.
** "Free time" is the complement of the number of working hours per week.

# Graphic constructions

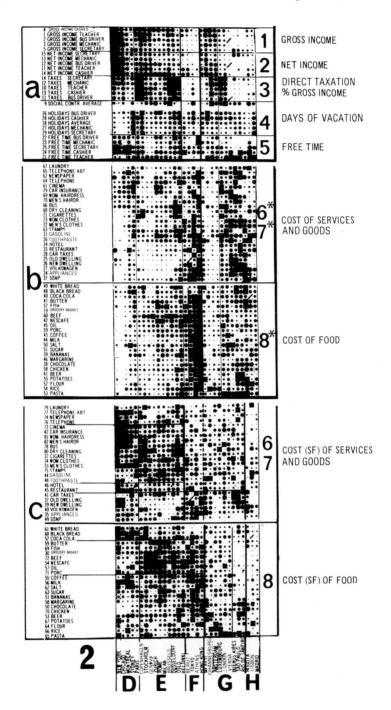

*The reorderable matrix* 47

*Classing n° 2* is based on prices in Swiss Francs (c). In which country should we have bought men's clothes in 1970? We should have gone to London or Luxembourg. But this was not as true for women's clothes.

For the international traveler and the banker, the price of goods and services (**6, 7**) divides the world into two main categories. Groups D (North America, Sydney and Paris) and E (Scandinavia, Switzerland, Italy, Belgium and Germany) are expensive. Group (F) with Beirut, Tokyo, Athens and Hong Kong is inexpensive for certain goods but expensive for food, particularly in Tokyo. Group (G) includes the least expensive cities and group (H) the ideal cities, except for some immediately noticeable items.

Classing n° 2 enables us to make the most of international travel and underlines the strong currencies in 1970.

Finally, note that the prices in Swiss Francs (c) and the real costs (b) have a very different distribution, which shows the danger of an anecdotal interpretation based solely on the initial data.

*Does graphic processing involve a great deal of time?*

Graphic processing is not instantaneous. Some people criticize it for this; but are they not confusing automation and analysis? These examples provide an answer:

— Is it better to spend three weeks graphically processing the information above or to publish meaningless anecdotes as the press did?

— An investigator presents his study, with accurate results. How much time did he spend on the calculations, aggregations and tables? *Eight months!* If it only takes three days to construct a matrix and three weeks to process and interpret it more deeply, can we really complain about the time spent on graphic processing?

— A study is comprised of fifteen factor analyses. How much time did it take to interpret them? *Six months!* The same result, but in a much more communicable form, can be obtained in less than a month by a combination of factor analysis and graphic matrix analysis.

Graphics is not a completely automated process. It involves all the stages of analysis, including communication. Accordingly it is fair to compare only what is comparable: the total time spent on a study.

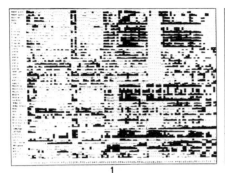 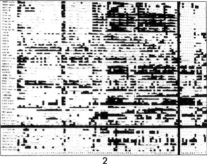

1    2

### SIBERIAN HYDRO-METEOROLOGY: INTERPOLATION MATRIX 100 × 52

In a recent study* Serge Bonin reviews 100 meteorological stations on the Euro-Asiatic coast of the Arctic Ocean about which we know fifty-two indicators defining various characteristics: rainfall, snow, hours of sunshine, nebulosity, wind, salinity of the sea, waves, ice, temperature, radiation, altitude. However, the data are incomplete for certain stations and indicators. The reorderable matrix enables us, among other results, *to reconstitute the missing data* with good probability. Here is the procedure followed.

The graphic transcription of the data table constructs our "zero" matrix **(1)**. The rows and columns which include unknown data are removed and placed respectively at the bottom and on the right **(2)**. The fully documented sections are then simplified without consideration of the removed portions. This simplification constructs a *homogeneous structure* **(3)**. In order to reintroduce the removed rows and columns into this structure we must rely upon *improbabilities*. For example the upper part of column ninety-five contains relatively large numbers. It cannot, therefore, be placed in the left-hand part of **(3)**. This means that we will focus on the right-hand part in seeking the two neighboring columns whose data are the closest to the known data in ninety-five. The position of ninety-five turns out to be between columns eighty-two and eighty-nine. In this way we can determine the probable place of each removed column in the homogeneous structure. The same applies to the rows. The final result is **(4)**.

The unknown data will probably have the same value as that of known neighboring data. This example is interesting because it may be verified. Soviet statistics offices subsequently

---

*Le traitement graphique d'une information hydro-météorologique relative à l'espace maritime du Nord Soviétique.* Paris: Mouton, 1974.

## The reorderable matrix

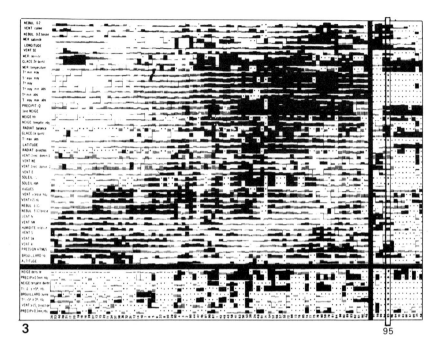

3

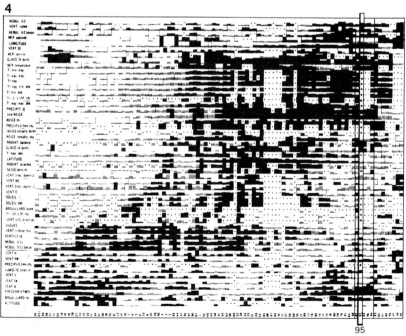

4

# Graphic constructions

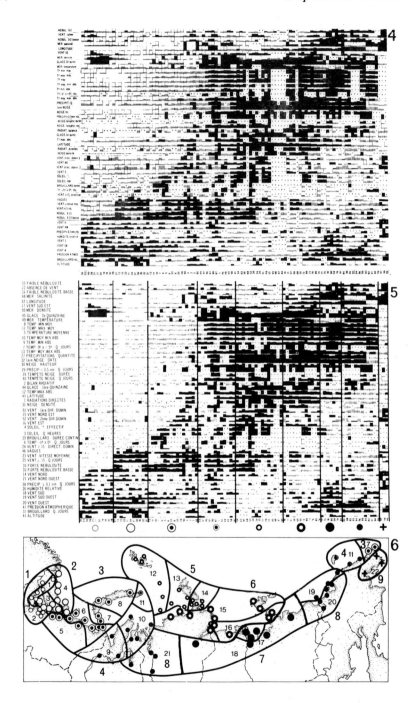

# The reorderable matrix

supplied a substantial part of the missing data, enabling us to construct matrix **(5)**, which has exactly the same order as **(4)**. The practicability of reconstituting unknown data by proximity and matrix interpolation is remarkably well confirmed here.

The probable accuracy of an interpolation depends: a) on the ratio of unknown to known cells; and b) on the regularity or irregularity of the characteristics' profiles. In the example on the facing page our chances are excellent since only 18% of the data is unknown and since meteorological phenomena generally produce regular profiles. The regionalization in **(6)** is both a revealing and a confirming of the relevance and interest of the typology **(5)**.

*RECREATIONAL PARK USAGE.*
*INTERPOLATION MATRIX 54 × 32*

Automatic highway meters record daily use of various access roads to a recreational park. Sixteen days of manual counting provide complete information on the use of each of the park facilities. Will the different highway meters alone enable us to predict the use of different facilities and thus provide a better daily distribution of services and personnel?

On $x$ the data table places a set of days including the sixteen which were manually counted. On $y$ it indicates the roads and recreation facilities. The chronological sequence of days is abandoned in favor of a typology. The "roads" matrix is simplified into three types of highway use: A, roads two, three, seven; C, roads five, six, four; B, scattered use. The sixteen manually counted days are distributed over these three types. A vertical reclassing of facilities separates those whose use tends to correspond with highway use A. They are different from those corresponding with B. On the other hand, there are only two counts in C: one Monday and one Tuesday. This is obviously insufficient. However, the two counts suggest that C corresponds to light and scattered use. Meteorological data, unfortunately missing, could certainly be added in some rows.

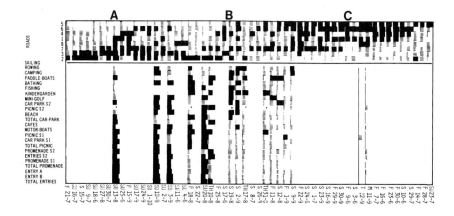

## Graphic constructions

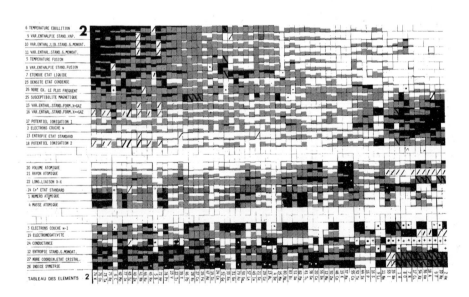

# The reorderable matrix

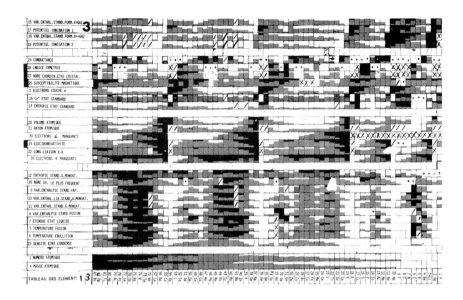

3. *MENDELEEV'S CLASSIFICATION.*
*PEDAGOGICAL MATRIX 83 × 27**

This is an example of the pedagogical use of reorderable matrices. We give a group of students an alphabetical list of the elements associated with twenty-seven characteristics **(1)**. The students are to simplify the matrix. Successive simplifications, such as **(2)**, for example, show that the set cannot be conceived as one unit. The structure of the upper part in **(2)** disorganizes the two lower parts. But these make the systemic relations appear. The students progressively discover the system organizing all the characteristics; in fact, they discover Mendeleev's periodic table **(3)**. This exercise thus becomes more than a learned lesson. Note that in **(3)** it would be possible to reconstitute a certain number of missing items with high probable accuracy.

\ / X: *undetermined values*
+ *and white dots: very small or very large values*

*From I. ARDITI, D. DEVEZE, D. MARBACH, *L'utilisation de la graphique dans l'enseignement supérieur de la chimie.* Université de Paris VII, Dépt. audio-visuel, 1973.

*THE IONIC CAPITAL. MATRIX 82 × 78\**

*Problem.* As evidence of a noteworthy civilization, Ionic capitals have long been the object of particular attention on the part of researchers. The capitals are not numerous and constitute a finite set of objects which probably will not be much augmented. The eighty-two pieces studied here represent the essential aspects of known capitals. Their study has led specialists to formulate typologies, and each has formulated his own, based subjectively on one or another characteristic.

\*From D. THEODORESCU. *Le chapiteau ionique.* Thèse: Université de Paris I, 1973.

## The reorderable matrix

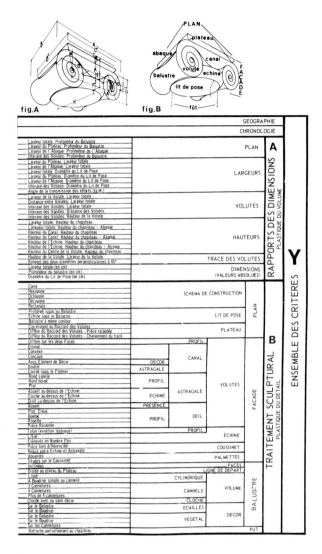

A more objective approach leads us to inventory all the possible variables, to consider them initially as equivalent characteristics, and to then apply them to the entire set of capitals. We then look for any apparent similarities that emerge during this analysis, either among the capitals or among the characteristics. Finally, we separate out the most representative characteristics of the groups thus obtained.

*The information and its graphic transcription.* The capitals are entered in $x$, the characteristics in $y$. Twenty-eight measurements involving ratios or dimensions (A) are taken for each capital and recorded on the matrix in nine degrees running from white to black. The qualitative model provides fifty yes/no indicators, recorded in (B). Missing data are represented by the symbol $(=)$, doubtful data by (M). The graphic transcription is accomplished by using the "Domino 1" apparatus.

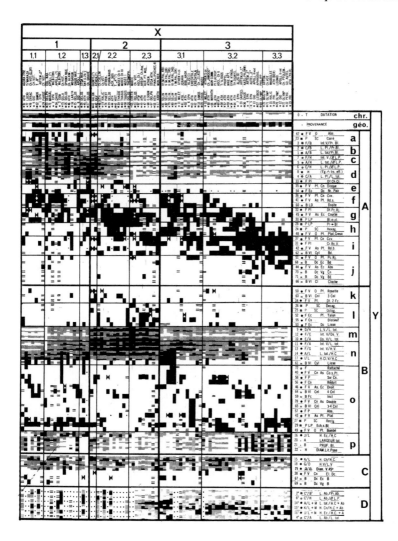

*Processing and interpreting:* By successive permutations, similarities emerge and lead to figure (**2**), which reveals three systems of relationships among characteristics along *y* and three main types of capitals along *x*. System (A) displays a series of characteristics which evolve chronologically (chr.). On the other hand, system B displays characteristics only present or important (k, l, m) at the middle of the chronological order or, conversely (o, p), at the extremities. Finally, characteristics C and D seem unrelated to chronology. These observations are schematized in an *"interpretation matrix"* (**3**). System A may be characterized by the design of the capital (b), which becomes less rectangular through time, and the shaft of the column (c), which becomes less narrow. System B can be characterized by the size of the volute (m), which is small in the beginning, then larger, then smaller again at the end of the period, while the overall size of the capital (p) undergoes an inverse evolution.

# The reorderable matrix

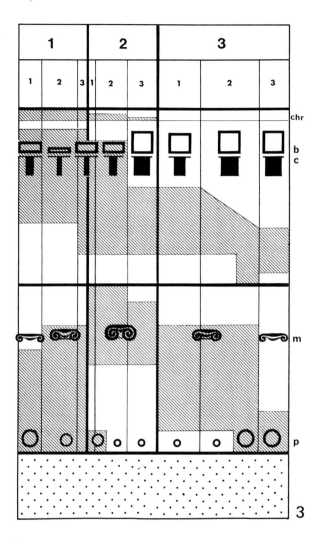

These two systems between them account for 85% of the characteristics and thus correctly represent the set of stylistic characteristics of the Ionic capital.

*Typology:* These results enable us to divide the capitals into three main types, each containing three sub-types. Their essential characteristics appear in each column of schema (3). Note in (2) that the chronological order (chr.) is not strict: certain capitals are out of place. The explanation is simple. It is obvious that at any given date a sculptor can be inspired by former examples or, on the contrary, be innovative. Consequently, the $x$ order of matrix (2) is that of style, not construction dates.
This typology can be verified. It is sufficient to rearrange the matrix according to the typology proposed by any given specialist. In no case does one of these subjective typologies account for more than 5% of the characteristics.

*"Scalogrammatics"*

Scalogrammatics is a procedure for interpretation by separation out, developed by V. Elisseeff in order to reconstitute a probable chronology of objects and characteristics, particularly in archaeology. The data must be comprised of yes/no items, and the method requires a double notation: the presence and absence of a characteristic. Consider* the sequence of changes taking place in the chronological evolution of ancient Chinese vases. Each object A, B, C... is defined by a set of characteristics one, two, three... Simplification, which can be automated, leads to matrix **(1)**, a *sub-perfect scalogram*.

In this matrix, various characteristics are disruptive, but it is possible that they are merely anecdotal. These characteristics are separated out until only one interruption per object remains. The figure obtained is a *near-perfect scalogram of the objects* **(2)**. The separation out of all the disruptive characteristics **(3)** constructs a *perfect scalogram of the objects*. Here it defines a linear sequence of objects and remaining characteristics.

Separation out in the other direction follows a similar course. It is possible that one disruptive object could interrupt an otherwise homogeneous set. For this reason **(4)** sets aside objects P and N, which display disruptions in more than half the cases. This separation out constructs a *near-perfect scalogram of the characteristics* **(4)**. Removal of all the disruptive objects constructs a *perfect scalogram of the characteristics* **(5)**.

This procedure reveals:
a) a linear sequence **(3)** of objects and characteristics which we can consider as being of a "chronological" nature.
b) disruptive characteristics — shape of the handle (seven-eight), presence of a ridge (nineten), presence of a neck (eleven-twelve) — which do not appear, in relation to the rest, to present signs of evolution.
c) disruptive objects which deserve closer study. The case of vases P and N is very significant; they disrupt four characteristics. Furthermore, they are the only ones possessing characteristic two, a transverse handle. This is thus a remarkable characteristic which leads us to create a sub-type, analysis of which shows it to be a continuation of a very unique example of the "zoomorphic" vases. Detailed analysis of these scalograms, to which must be added a number of objects of each variety (F and H for example are only single specimens), enables us to evolve a chronological hypothesis **(6)** involving major periods separated by a transitional style.

Scalogrammatics highlights special cases and measures the extent of their particularity by comparison with a homogeneous structure. An interpretation of the balance between these two sets enables us to formulate a hypothetical chronology that excludes none of the elements likely to contradict it. This method can be applied to any hierarchical set.

---

*From V. ELISSEEFF. "Application des propriétés du scalogramme à l'étude des objets," *Journées d'Etudes sur les Méthodes de Calcul dans les Sciences de l'Homme*. Paris: CNRS, 1968; and *Archéologie et calculateurs, Problèmes sémiologiques et mathématiques*. Marseille: CNRS, 1970.

# The reorderable matrix

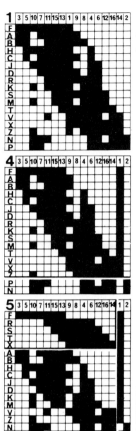

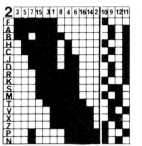

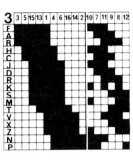

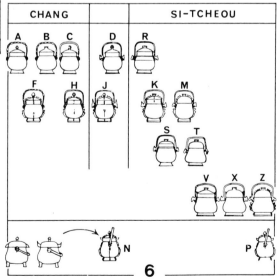

## B.2.2. THE WEIGHTED MATRIX

In the reorderable matrix the elementary areas are equal. In the weighted matrix $x$ and $y$ vary in relation to a certain quantity. The *areas become meaningful*; the rows and/or columns are unequal. The weighted matrix must therefore be drawn and can only be applied to tables of limited dimensions.

### B.2.2.1. Principle of construction

1st example. Take table **(1)**. It shows the distribution of twenty-five individuals according to three professions A, B, C and three regions *n, c, s*.
(2) Calculate the vertical percentages.
(3) Construct drawing **(3)** directly according to these percentages. Darken whatever exceeds the mean per row. Reclass rows and columns (for purposes of simplification, a pre-classed example has been chosen here).
(4) Give the columns a width proportional to the totals obtained from **(1)**.
(5) In the final drawing, write the totals per column. The weighted matrix **(5)** shows:
- the totals by profession along $x$,
- the percentage of each profession in each region along $y$,
- the partial quantities by area,
- and whatever exceeds the mean in black, that is, whatever characterizes each region and each profession.

2nd example. Take the data in **(6)**. They apportion the trade of the seven COMECON* countries over the different economic blocs* for 1968. The table has limited dimensions.
Note that the table is in percentage per column, i.e., the information is already at the second stage of construction, as compared to the previous example. We may now:
- construct **(7)** directly according to the percentages;
- darken whatever exceeds the mean and re-class **(8)**;
- we have the totals per country in millions of dollars **(9)**;
- and give each column **(10)** a width proportional to these totals, which enables us to restore all the quantities graphically and to portray the true picture. For example, Bulgaria and the USSR, represented with the same volume of exchange in **(8)**, are given their respective shares in **(10)**, that is, ten and eighty-four million dollars.

## The weighted matrix

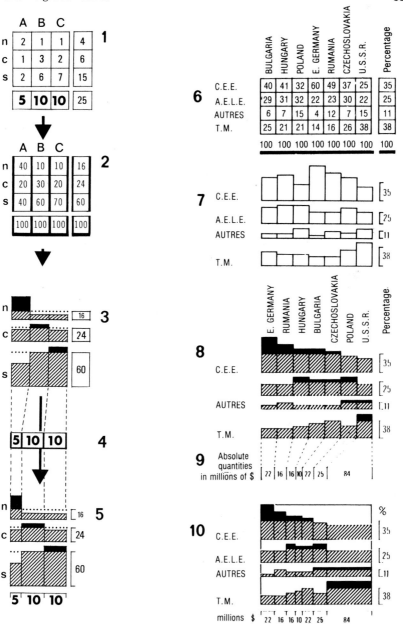

*COMECON — Council for Mutual Economic Assistance
CEE — European Economic Community (Common Market)
AELE — European Free Trade Association
AUTRES — Other countries outside COMECON
T.M. — Third world

### B.2.2.2. Application of the weighted matrix.

The preceding examples show that the weighted matrix is applicable to a data table whose row totals and column totals *are meaningful*. They are tables of the form zQob, that is, tables that apportion a single series of objects, portrayed in *z*, across two characteristics, portrayed respectively in *x* and *y*. We shall see on page 255 that the zQob tables contrast with the zD and z01 tables in which *x* represents the objects and *y* the characteristics and in which the totals are not meaningful. In such cases, the weighted matrix is not applicable.

### B.2.2.3. — Meaning of the weighted matrix.

Let us return to example **(1)** on page 60. The percentages calculated in the columns result in **(4)**. Calculated in rows, they result in **(10)**. These two constructions have exactly the same meaning. In a table of the form zQob, the direction of 100 is unimportant (p. 249).

In fact, a weighted matrix portrays, cell by cell, the difference between the *observed distribution* **(1)** and the *expected distribution* **(6)**\* such as it results from the totals, that is, from all the quantities.

Consider the northern (n) and southern (s) regions in table **(1)**. There is a population almost four times greater in the south (fifteen individuals) than in the north (four individuals). Therefore it is probable that in each profession there will be four times more individuals in the south. True or false? This is the aim of the statistical study. Yet we observe that it is false in almost every case. There is much more A in the north than expected, likewise more C in the south. How do we determine our expectations? Simply by constructing diagram **(6)** in which column A represents five individuals, and columns B and C, *twice the width*, represent ten. Likewise, the widths of the rows represent four, six, and fifteen individuals, respectively. Consequently, the area of each cell represents the probable proportion of individuals in each case considered. An easy proportion to calculate. Each area is the product of the corresponding totals. We divide by the general total, twenty-five, in order to reconstitute numbers comparable to those of the elementary data **(1)**.

If we separate these proportional cells **(7)**, we then see **(5** and **11)** that the *weighted matrix* simply makes the difference between the expected

---

\*Also called "observed frequencies" and "expected frequencies."

## The weighted matrix

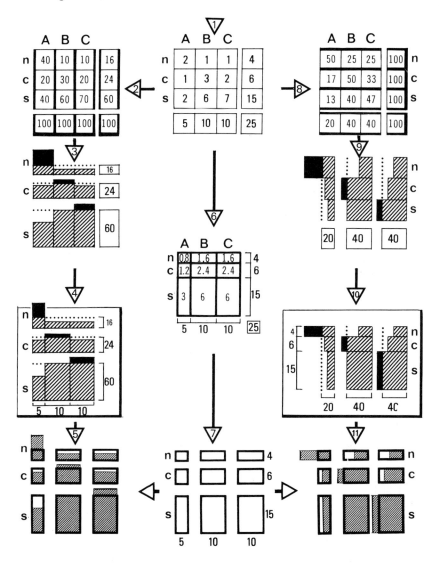

distribution (6) and the observed distribution (1) apparent. This difference is expressed vertically in **(4)** and horizontally in **(10)**.

*The direction of 100 being unimportant in weighted matrices,* it is preferable to choose the direction which leads to the squarest construction, which amounts to a preference for the shortest additions and therefore the greatest number of "100s."

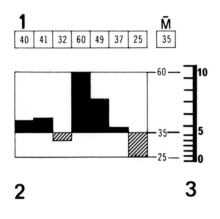
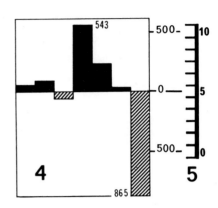
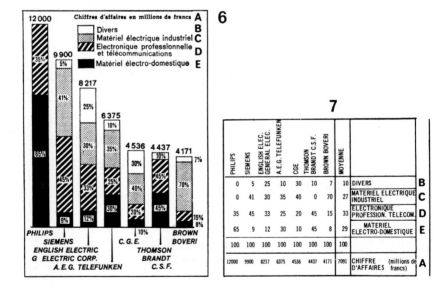

A. Annual volume, B. Miscellaneous, C. Electric industrial equipment, D. Electronics and telecommunications, E. Electric household appliances.

*The weighted matrix* 65

### B.2.2.4.  Reclassing a weighted matrix

A weighted matrix of limited dimensions can be reclassed on sight. In the COMECON example from p. 60, it is easy to reclass figure **(7)** simply by careful study and a few sketches. For larger dimensions, *it is necessary to first transcribe the information into a reorderable matrix.* Which numbers should be transcribed and what sort of step series should be applied?

*From the percentage table.* In a percentage table, deviations from the mean are not weighted. Take the first row **(1)** in the COMECON table on p. 60. Diagram **(2)** represents it. Diagram **(4)** portrays the weighted deviations (multiplied by the total of each column). The application in **(5)** of a range of steps which takes extremes into account is the best (see p. 197), but it entails calculation of the weightings **(4)**. We see that the application in **(3)** is nearly equivalent. *It places the middle step on the mean* and can treat the spacing of the steps above and below separately, if necessary. Nevertheless, extremes which are too great should be removed. This arrangement is only valid for weighted matrices.

*From the absolute-quantity table.* Here it is advisable to calculate the algebraic deviation $X$ (chi)* between the observed distribution (0) and the expected distribution (E).

- Cell by cell, E is the product of the corresponding marginal sums, divided by the total sum.

- Cell by cell, $X = \dfrac{O - E}{E}$. It can be positive, nil or negative.

It is the $X$ table which is set up as a reorderable matrix. The nil $X$ must correspond with the middle step of the visual scale.

B.2.2.5. — EXAMPLES

*EUROPEAN ELECTRICAL INDUSTRY IN 1969.*
*WEIGHTED MATRIX 7 × 4\*\**

Take diagram **(6)** which appeared in an influential daily newspaper. It divides a set of annual volume figures into two reorderable variables: different companies and different products. It gives the absolute quantities per column. Its dimensions are limited: the standard construction is a weighted matrix. After having reconstituted the data table **(7)**, the dimen-

*The author's use of the notation $X$ is somewhat unorthodox and should not be confused with the conventional $X^2$ statistic, from which it differs subtly (Editor's note).
\*\*From J.D. GRONOFF. "Cartes et diagrammes dans la presse," *Communication et langages,* Paris, 1973.

*Graphic constructions*

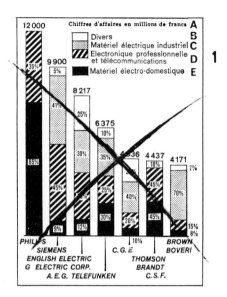

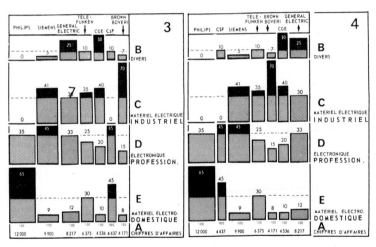

sions of the table enable us to draw the weighted matrix **(3)** directly, to cut it apart, and to reclass it **(4)**. The amounts of black above the mean indicate each company's specialization. *Interpretation.* This example effectively underlines the difference between *simplification,* the passage from **(3)** to **(4)**, and *interpretation*, which can lead to different classings, each responding to a precise question.

**(5)** isolates the two French companies CSF and CGE. It shows that they are complementary. It also shows that each is faced with competition from much larger companies in every branch.

*The weighted matrix*

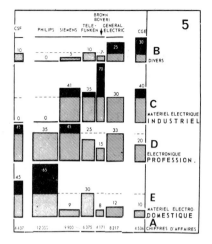

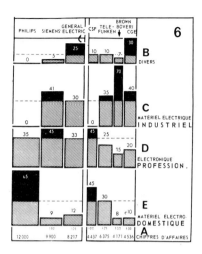

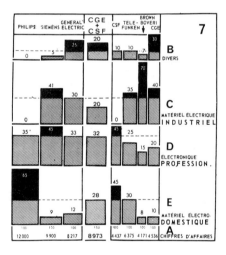

**(6)** separates the companies into two groups: the larger (above the mean) and the smaller. We thus discover that in each group the companies divide activities in a harmonious way that tends to limit competition.

**(7)** constructs the fictitious company which, according to the newspaper article, would result from the amalgamation of the two French companies. It would be above the mean and belong to the group of "large" companies. It would be similar to the British General Electric Company. But it would disrupt the harmony which characterizes the existing structure. Unfortunately, and because of construction **(1)**, these observations, except for the total size, were not perceived by the journalist.

# Graphic constructions

| Catégorie socio-professionnelle | Lecture de quotidiens (1) | Lecture d'hebdomadaires (2) | Lecture de mensuels, trimestriels (3) | Temps passé à la lecture de livres, journaux (4) | Lecture récente d'un livre (5) | Lecture de livres (6) | Cinéma (7) | Théâtre (8) | Variétés (9) | Concerts (10) | Château, monument (11) | Musées d'art (12) | Salon ou foire exposition (13) | Sorties au café (14) | Sorties au restaurant (15) | Réceptions de parents et d'amis (16) | Sorties le soir (17) | Invitation de parents et d'amis (18) | Participation à des associations (19) | Entretien du véhicule (20) | Jardinage (21) | Bricolage (22) | Télévision (23) |
|---|---|---|---|---|---|---|---|---|---|---|---|---|---|---|---|---|---|---|---|---|---|---|---|
| Salariés agricoles | 39,3 | 26,2 | 14,5 | 41,6 | 33,9 | 10,4 | 10,6 | 8,7 | 22,0 | 3,0 | 18,3 | 9,9 | 26,8 | 24,9 | 17,5 | 32,3 | 28,5 | 34,9 | 12,5 | 51,1 | 55,1 | 38,2 | 46,9 |
| Agriculteurs exploitants | 53,9 | 50,9 | 25,3 | 55,6 | 34,9 | 13,7 | 16,2 | 9,6 | 15,1 | 4,1 | 18,3 | 3,9 | 30,3 | 35,5 | 9,1 | 30,6 | 22,5 | 20,5 | 19,9 | 40,0 | 70,6 | 19,7 | 40,2 |
| Personnels de service | 49,7 | 37,3 | 19,1 | 54,3 | 29,0 | 33,6 | 19,1 | 23,6 | 4,3 | 25,9 | 15,4 | 44,1 | 25,0 | 25,2 | 31,9 | 31,6 | 13,5 | 26,9 | 20,7 | 29,9 | 52,2 | | |
| Ouvriers | 56,0 | 47,3 | 20,8 | 60,5 | 51,9 | 27,9 | 28,6 | 12,3 | 19,5 | 4,3 | 27,2 | 12,5 | 30,8 | 36,1 | 24,0 | 36,3 | 29,4 | 33,1 | 11,5 | 40,1 | 42,6 | 41,5 | 54,4 |
| Employés | 60,8 | 47,9 | 30,2 | 72,0 | 69,4 | 42,1 | 31,0 | 27,7 | 25,9 | 11,2 | 36,3 | 21,0 | 42,2 | 30,7 | 40,2 | 44,9 | 36,1 | 43,7 | 17,3 | 31,8 | 23,7 | 34,9 | 55,7 |
| Professions indépendantes | 71,3 | 54,3 | 35,2 | 73,5 | 66,6 | 36,6 | 34,2 | 31,5 | 25,8 | 8,6 | 38,3 | 20,7 | 44,0 | 40,4 | 42,4 | 46,9 | 35,3 | -1,9 | 17,0 | 38,0 | 36,3 | 35,8 | 61,2 |
| Cadres moyens | 62,3 | 59,7 | 47,3 | 75,7 | 88,2 | 55,8 | 40,0 | 42,5 | 35,7 | 18,6 | 55,8 | 38,2 | 49,5 | 32,9 | 54,2 | 52,0 | 45,3 | 55,1 | 26,6 | 38,8 | 35,0 | 41,4 | 54,6 |
| Cadres supérieurs | 65,2 | 72,7 | 57,7 | 83,5 | 92,9 | 71,6 | 56,4 | 59,9 | 31,5 | 30,5 | 66,0 | 54,5 | 48,2 | 31,9 | 65,8 | 65,6 | 67,6 | 52,2 | 66,6 | 34,7 | 24,4 | 32,6 | 40,4 | 46,7 |
| Autres actifs | 74,4 | 69,1 | 33,9 | 84,1 | 85,3 | 47,1 | 26,9 | 31,9 | 25,2 | 13,0 | 39,0 | 23,0 | 46,2 | 23,4 | 27,8 | 38,1 | 31,7 | 36,2 | 32,9 | 35,8 | 34,0 | 63,5 | |
| Inactifs | 60,4 | 40,5 | 26,2 | 68,2 | 44,9 | 25,4 | 18,7 | 13,6 | 13,0 | 6,9 | 18,5 | 13,1 | 19,6 | 22,8 | 19,1 | 29,4 | 19,9 | 30,6 | 11,4 | 23,4 | 48,1 | 26,8 | 45,1 |
| **Revenu annuel du ménage** | | | | | | | | | | | | | | | | | | | | | | | |
| Moins de 6 000 F | 49,8 | 29,6 | 15,6 | 56,6 | 31,3 | 15,6 | 9,3 | 6,2 | 6,8 | 1,7 | 9,3 | 5,5 | 15,1 | 16,9 | 7,9 | 18,6 | 13,0 | 26,3 | 8,8 | 21,6 | 51,3 | 21,2 | 29,1 |
| De 6 000 à 10 000 F | 53,5 | 38,0 | 22,0 | 58,5 | 43,0 | 21,5 | 20,7 | 12,3 | 17,1 | 6,6 | 20,9 | 10,5 | 22,4 | 25,7 | 16,5 | 32,8 | 23,0 | 32,7 | 11,6 | 35,4 | 48,5 | 34,8 | 43,7 |
| De 10 000 à 15 000 F | 57,8 | 46,9 | 26,5 | 65,1 | 50,3 | 28,7 | 27,1 | 14,5 | 20,5 | 6,2 | 27,4 | 13,9 | 32,1 | 35,7 | 24,4 | 34,8 | 30,5 | 33,5 | 15,8 | 38,6 | 45,7 | 38,9 | 53,3 |
| De 15 000 à 20 000 F | 59,9 | 50,5 | 28,8 | 65,8 | 60,6 | 35,5 | 31,1 | 19,6 | 21,1 | 6,8 | 31,6 | 17,3 | 38,9 | 37,6 | 30,9 | 40,6 | 33,8 | 37,4 | 16,1 | 38,5 | 41,2 | 39,4 | 58,7 |
| 20 000 F et plus | 68,3 | 61,2 | 40,9 | 77,5 | 75,2 | 48,8 | 39,9 | 38,8 | 29,3 | 16,4 | 46,7 | 31,8 | 45,4 | 34,0 | 48,7 | 53,8 | 29,9 | 45,7 | 22,5 | 33,2 | 35,8 | 40,2 | 58,7 |
| **Niveau d'études** | | | | | | | | | | | | | | | | | | | | | | | |
| Pas de diplôme | 52,2 | 36,3 | 16,3 | 56,1 | 36,2 | 18,1 | 17,9 | 10,2 | 13,7 | 3,7 | | | | | | | | | | | | | |
| Certificat d'études primaires | 63,5 | 49,8 | 28,5 | 68,7 | 58,1 | 28,5 | 24,8 | 16,0 | 19,8 | 5,3 | | | | | | | | | | | | | |
| Brevet | 64,8 | 59,4 | 40,9 | 75,4 | 81,4 | 50,5 | 44,5 | 34,2 | 33,1 | 15,0 | | | | | | | | | | | | | |
| Baccalauréat, et études supér. | 66,1 | 67,7 | 57,4 | 82,4 | 52,8 | 72,3 | 51,7 | 57,0 | 31,9 | 27,7 | | | | | | | | | | | | | |
| **Sexe** | | | | | | | | | | | | | | | | | | | | | | | |
| Hommes | 63,5 | 45,8 | 30,4 | 67,8 | 57,9 | 34,4 | 30,8 | 21,4 | 24,2 | 9,1 | | | | | | | | | | | | | |
| Femmes | 56,4 | 50,0 | 27,3 | 65,1 | 56,3 | 30,7 | 25,5 | 20,3 | 17,9 | 7,8 | | | | | | | | | | | | | |
| Ensemble | 59,7 | 48,0 | 28,8 | 66,3 | 57,0 | 32,5 | 28,0 | 20,8 | 20,8 | 8,5 | | | | | | | | | | | | | |
| **Age** | | | | | | | | | | | | | | | | | | | | | | | |
| 14 à 24 ans | 50,3 | 53,3 | 30,9 | 60,2 | 81,8 | 57,1 | 67,6 | 28,5 | 38,0 | 11,0 | | | | | | | | | | | | | |
| 25 à 39 ans | 55,8 | 51,0 | 31,9 | 64,4 | 61,4 | 33,3 | 28,5 | 25,9 | 26,0 | 10,1 | | | | | | | | | | | | | |
| 40 à 59 ans | 65,6 | 48,1 | 29,6 | 59,8 | 53,5 | 27,1 | 27,7 | 20,2 | 17,0 | 9,6 | | | | | | | | | | | | | |
| 60 ans et plus | 64,1 | 40,7 | 23,0 | 69,2 | 36,5 | 17,6 | 6,7 | 10,5 | 6,5 | 4,8 | | | | | | | | | | | | | |

Source: enquête sur les comportements de loisirs de 1967.

**1**

| | |
|---|---|
| 10 CONCERT | 104 |
| 19 ASSOCIATIONS | 190 |
| 12 MUSEUMS | 212 |
| 9 MUSICAL ENTERTAIN | 237 |
| 8 THEATRE | 257 |
| 7 CINEMA | 296 |
| 14 CAFE | 303 |
| 3 MONTHLY MAGAZINES | 310 |
| 15 RESTAURANT | 325 |
| 17 EVENING OUTINGS | 332 |
| 11 SIGHT-SEEING | 344 |
| 20 VEHICLE | 347 |
| 6 BOOKS | 359 |
| 22 HANDICRAFT | 363 |
| 13 SHOW, FAIR | 382 |
| 18 INVITATIONS | 394 |
| 21 GARDENING | 400 |
| 16 RECEPTIONS | 410 |
| 2 WEEKLY NEWSPAPERS | 502 |
| 23 TELEVISION | 520 |
| 1 DAILY NEWSPAPER | 593 |
| 5 RECENT BOOK | 662 |
| 4 READING | 699 |
| OVERALL | 8 471 |

**2**

S.P.C. INCOME STUDIES AGE SEX

*STATISTICAL DATA ON LEISURE.*
*WEIGHTED MATRIX 23 × 25*

A double page (1) from a statistical study* itemizes replies to a questionnaire on leisure. These replies are divided according to 1) Socio-professional categories, 2) Income, 3) Educational level, 4) Age, 5) Sex. Each cell gives the number of positive replies per 100 persons in the category considered.

It is not sufficient to represent these numbers and to class them. Diagram (2) proves this. It classes the leisure activities from top to bottom according to use by the overall population. In each table it classes the categories from left to right according to the total use in each col-

*Données sociales,* première édition. Paris: INSEE, 1973.

*The weighted matrix* 69

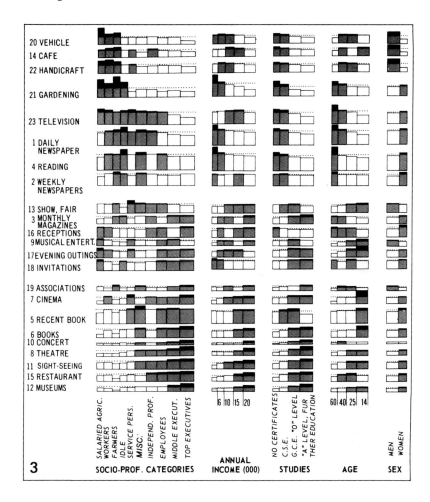

umn. Essentially, everything is compatible and compatible with the totals. Looking more closely we discover a few irregularities. But, to compare them! Yet, it is precisely the comparison of these "irregularities" which justifies the cost of each survey. It is therefore necessary to "level" the variation resulting from the totals and to make the "irregularities" appear.

This is what the weighted matrix (3) will show. It transcribes the overall variations along $x$ and displays along $y$ the irregularities, that is the individual tendencies. We highlight them here by leaving those which are below the mean in "white." The weighting leads to a reclassing of socioprofessional categories and leisure activities. Incidentally, this classing is highly significant for the other components and *enables us to compare them with each other.* Note in passing that the other components are ordered along $x$ and thus correspond to image-files.

*B.2.3. — THE IMAGE-FILE*

This is the standard construction for data tables ≠ 0, that is, where one component is ordered, as with the hotel example.

### B.2.3.1. — Permutation devices

*Cards laid flat* (see example page 4). We compose a "key" (or legend) card describing the ordered component, then one card per row of the table. The cards are then lined up along a sheet of paper folded about one cm from the side. On the right, we leave sufficient space to write the card's title. On the left and the right we leave an empty margin about five cm wide to fix the cards with two-sided adhesive tape each time that a classing is photocopied. This formula is usable for up to about thirty rows. It enables us to enter the quantities along $y$ and consequently to compress the columns, which can be numerous. Darken the largest numbers, that is, usually those above the mean.

*Cards on edge.* Strips of cardboard one mm thick and four cms high can produce a very convenient image-file **(1)**. But sheets of paper 21 × 29.7 folded into four **(2)** are just as convenient and lighter. The fold should not be creased.
A survey sheet can be used directly (page 89). The cards or sheets are stacked to the left along a guide-bar fixed to a board **(3)**. The key (legend) is mounted on an appropriate border **(4)**. The quantitative columns are generally fifteen mm wide **(5)**. The yes-no columns should not be less than two mm **(7)**.
A precise series of dates can be transcribed as in **(8)**.
Eliminate all double notations. For example, only one column is needed **(12)** to indicate the male-female characteristic **(9)**.

To perceptibly improve the legibility of the image-file and the matrix-file (p. 86):

1 - In large quantitative series, take the mean or any average quantity as an axis and darken everything above it **(6)**; record everything below it using well separated vertical strokes (p. 229).
2 - In a matrix-file, class the yes-no columns insofar as possible by ordered categories and construct these categories into a visually ordered column. Thus the categories — "savanna," "forest," "bare ground,"

*The image-file*                                                                71

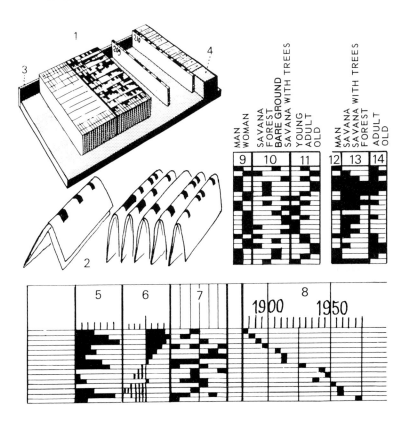

"savanna with trees" **(10)** — should be ordered: bare ground, savanna, savanna with trees, forest. The bare ground is represented by white, the forest by black extended over three columns **(13)**. The four columns of **(10)** become a single column, ordered in terms of plant cover **(13)**. The same is true for the age categories in **(11)** and **(14)**. The second file is much more legible and smaller.

*The "Domino" apparatus* used for matrices is all the more applicable to image-files. Titles can be written on the folded edge. *The cathode screen* can display an image-file and be used to carry out the key-board permutations.

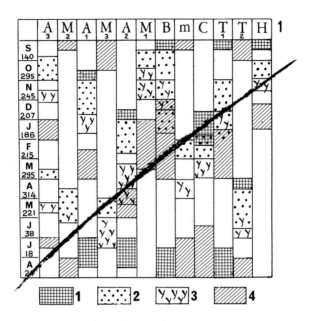

## B.2.3.2. EXAMPLES

### CROP CYCLES IN AN AFRICAN REGION. IMAGE-FILE 13 × 17

The data are given in figure **(1)**: from left to right the different crops*, from top to bottom months and monthly rainfall. No groupings emerge from this figure. It must be read entry by entry.

*Construction of a "flat" image-file.* The legend **(2)** records the months twice, in order to study potential periods, and it displays the rainfall (P) graphically. For each crop we compose a separate card on a sheet 21 × 29.7 **(3)**. The notations **(3)** have a height of one cm. The various types of work** are differentiated by a series of values which run from light: ground-breaking, to black: harvesting **(4)**. The initial or "zero" image-file **(5)** reproduces the original classing **(1)**.

*Overall classing.* A first classing reveals two main periods. To confirm them we re-draw in duplicate the cards for the crops which precede and follow the grouping. They show clearly **(6)** that the most significant breaks do indeed occur between rows four and twelve and in mid-September.

| * A peanuts | H beans | **1 ground-breaking |
|---|---|---|
| M manioc | C squash | 2 sowing |
| m maize | T tobacco | 3 hoeing |
|  |  | 4 harvesting |

P rainfall
D in black: large distance between settlement and fields.

# The image-file

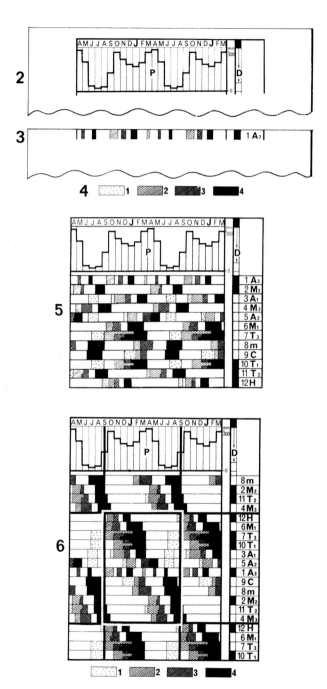

*Graphic constructions*

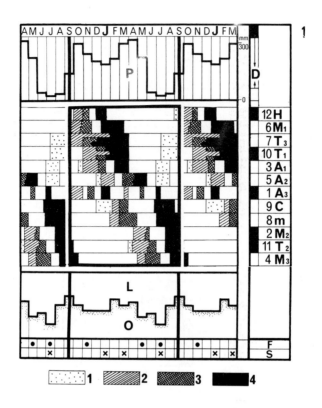

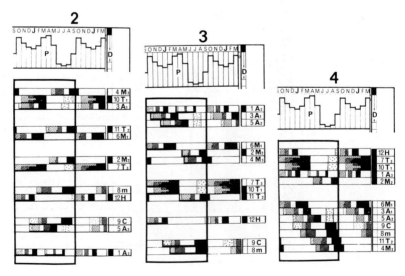

# The image-file

The crop cycles thus form image (**1**) to which we may then add various data: month by month we can total working time (O) and its complement: free time (L). Additional information, such as the occasion of major feasts (F) and market days (S), then acquires its full meaning. The periods of ground-breaking, hoeing, and harvesting are shown to be intimately linked to rainfall.

*Intermediate questions.* How are the crop cycles completed over time? Classing (**2**) provides the answer and reveals a very strict organization, taking migrations (D) into account. What are the characteristics of different varieties of the same crop? The answer is given in classing (**3**). The two periods of A3 emerge clearly. What are the characteristics of the crops in terms of their distance from the settlements (D)? Classing (**4**) shows that in the dry period crops are generally nearer the settlements.
This example underlines the experimental role of graphics.

## STUDY OF ANIMAL BEHAVIOR.
## IMAGE-FILE 96 × 12

This example is highly significant in the fields of data analysis and processing techniques.

Let us study the effect of light on animal behavior. How much time elapses before a wood-louse goes from light into darkness? Three compartments are placed one after another: I, direct light; II, shade; III, dark. Eight wood-lice are placed in I. Every five minutes during one hour, readings are taken of the compartment occupied by each animal. The experiment is performed twelve times.

*Data.* An apportionment (see page 235) produces the following table:
    Compartments (0)      3   x
    Wood-lice ($\neq$)    8   x
    Times (0)            12   x
    Experiments (0)      12   x

The problem is to reduce each component. In fact, what matters is to see:
1 - if all the wood-lice behave in the same way or if they form groups,
2 - if all the times are homogeneous or if certain periods are more significant,
3 - if all the experiments are comparable or if a variation in general conditions disturbs them.

The "compartment" component is not reducible since in fact it only includes two categories: the time spent in n°1 and that spent in n°2.

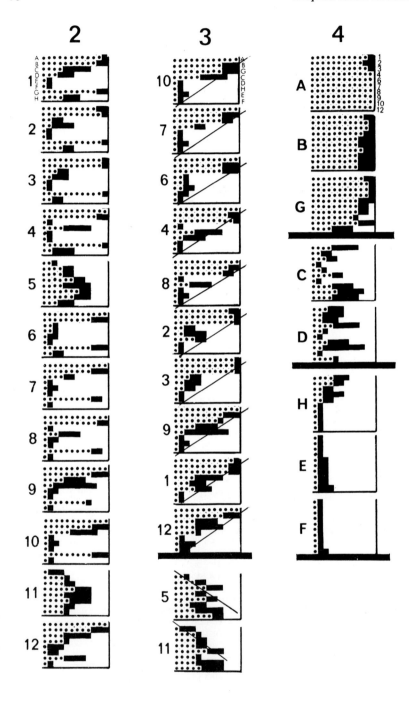

# The image-file

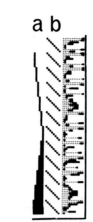

*Graphic construction.*
What do we know about one animal and one experiment? A total time taken to reach darkness, divided into two parts: t1 and t2 or to simplify:

        T.   Q total time        Q
        C.   Animals            8   x
        E.   Experiments    12  x

To discover potential groupings, the three components must be represented on the plane. However, the time, T, is ordered. We are therefore faced with an image-file placing time T along $x$ and the animals and experiments along $y$. It is sufficient to separate t1 and t2 by a visual difference: gray and black, for example. This file was set up on "Domino 3" equipment. In the file, the experiments are recorded as in (1a) and the animals as in (1b).

*Permutations and Interpretation.*
The zero classing corresponds to the actual sequence of the twelve experiments (2).

*The reduction of the component "twelve experiments"* is carried out in two stages:
a) The zero classing constructs one image per experiment. But we observe that generally animals A, B and G are very slow whereas E and F are fast. In each experiment we reordered the wood-lice A B G C D H E F (3).
b) This classing gives a diminishing form to all the experiments except five and eleven. The experiments thereby form *two groups*, and their difference can only correspond to a change in experimental conditions. Consequently, we exclude experiments five and eleven, which are not homogeneous with the majority (3), and we study the conditions under which they were carried out.

*Reduction of the component "eight wood-lice"* is also accomplished in two stages:
a) We construct one image per wood-louse, classed in order of experiments: 1, 2, 3, 4, 6 . . . excluding 5 and 11.
b) Then we class the wood-lice by similarity (4). They form three groups: A, B, G are slow; H, E, and F are fast; whereas C and D are "erratic." We can define each type of behavior and observe, for example, that A, C, E, and F tend to "tire" and that, on the contrary, B, G, and H learn to play the game. The wood-lice are far from being homogeneous.

# Graphic constructions

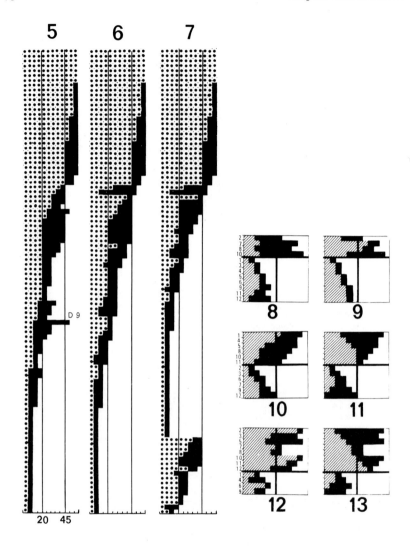

*Reduction of the "time" component* is simple. It is sufficient to class the times from the longest to the shortest. In **(5)** and **(6)** experiments five and eleven are included. Three times seem to be particularly significant: ten minutes, the maximum time for the fastest; twenty minutes, the time frequently taken to pass into compartment II; and forty-five minutes, the time often taken to pass into compartment II or into compartment III.

When we exclude experiments five and eleven **(7)** these times are confirmed. Time is therefore not homogeneous. There are three predominant times. Furthermore, the heterogeneity of the experiments does not depend on time. Finally, wood-louse D, in the ninth experiment, displayed exceptional behavior, which stresses the fact that all the others have a general tendency, expressed by the mean ratio of the times spent in each compartment.

## The image-file 79

*Permutations offer other possibilities.* To study an "erratic" wood-louse we could, for example, divide the experiments into "long" times and "short" times. If we maintain the order of the experiments in each group, observations can become very precise and combine long or short times, fatigue or training, compartment I or II. We see, for example, that **(12)** is truly "erratic", whereas **(10)** can be analyzed: in the long times he learns, whereas in the short times he tires. The total performances of **(11)** are comparable with those of **(10)**, but their behavior is completely different, etc.

What is particularly important to remember from this example is the following observation: when the components of a problem are few (fewer than about six), the goal of processing can be the successive reduction of each component as a function of the entire set of components. We construct a *collection of experiments* in order to reduce the experiments, a *collection of wood-lice* in order to reduce the wood-lice, a *collection of times* in order to reduce the times, etc.

The single drawing, constructed once and for all **(14)**, which supplied the data for this example, only exploits a tiny portion of the possibilities of graphics. But this is not a problem of graphics. It is simply because the data table was poorly constructed. A differential component ( ≠ ) such as "different animals" should never be placed in *z*. *A component ≠ must always be along x or y. We must be able to process it.*

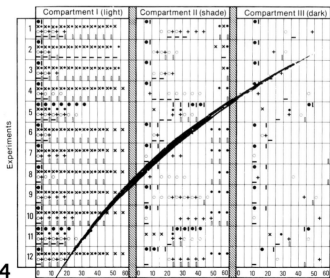

# Graphic constructions

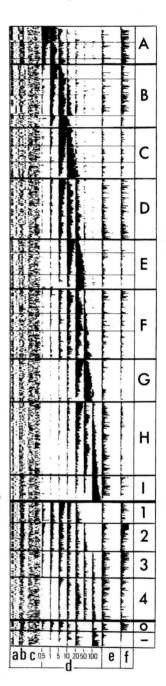

*The image-file* 81

*THE SIZE OF FARMING OPERATIONS IN PROVENCE.*
*IMAGE-FILE 1000 × 9*

*Data.* For each township we know* the percentage of area occupied by farms of less than one hectare, from one to five hectares, from five to ten hectares, etc. We also know the total useful area per township and the number of farms. There are 1000 townships. The data table has the form of 1000 × 7 ordered categories, plus two complementary items. The graphic construction is therefore an image-file with ($\neq$) 1000 along $y$.

*The key (or legend)* **(1)**. On the left it denotes (a) the county's, b) the canton's, and c) the township's numbers. In fact we will only use the county's number. In the center it lists percentages. Note that the seven columns are equal, even though the maximum percentages of area per category are 10% for the category less than 1 hectare, and 100% for the category more than 100 hectares. Such differences lead us to adopt an individual or proper scale for each category (see p. 211), without which a typology becomes invisible and impossible. Note also that it is the small categories which concern the greatest number of people. On the right we record the total area and the total number of farmers.
This file, constructed on cardboard cards 1 mm thick, was accordingly one metre long.

*Processing.* The initial or zero file follows the order of the statistical numbering **(1)**. After a first trial classing we observe on the one hand townships in which all categories are represented, and on the other townships clearly typed by two or three categories. The set thus divides into two systems: typed and non-typed townships **(2)**. The two systems are classed separately from the smallest to the largest farms and define types A, B, C . . . and 1, 2, 3 . . . Several townships, noted 0 and −, are characterized by the absence of the category 50-100 hectares. The resulting image **(2)** acquires its full meaning only when mapped.

*From Geneviève PEUGNIEZ. Faculté d'Aix-en-Provence.
a. County, b. Canton, c. Township, d. Size of farms (in hectares), e. Useful agricultural area, f. Number of farms.

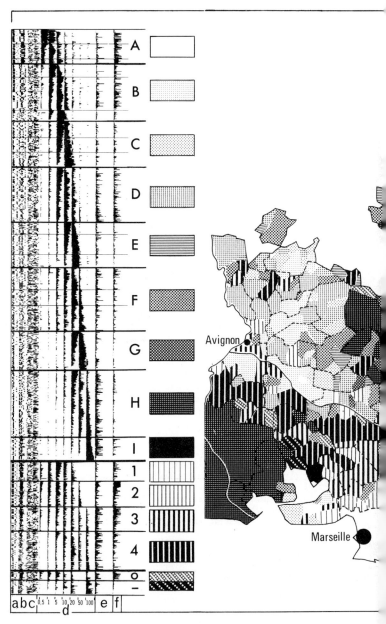

*Cartographic transcription.* In order to indicate the types on a map, only the 3rd, *z*, dimension of the image may be used. However, these types are ordered; the transcription may accordingly employ a variation of value, ordered from white to black. The two systems are differentiated by the fineness of the screens, that is, by their "textures," very fine for A, B,

## The image-file

C...and easily visible in the form of vertical lines for 1, 2, 3....The file itself constitutes the legend of the map, which, by the groupings it constructs, confirms the types and the systems in terms of their likelihood and usefulness. Within this regional framework, the groupings discovered become comparable with any other mapped information.

# Graphic constructions

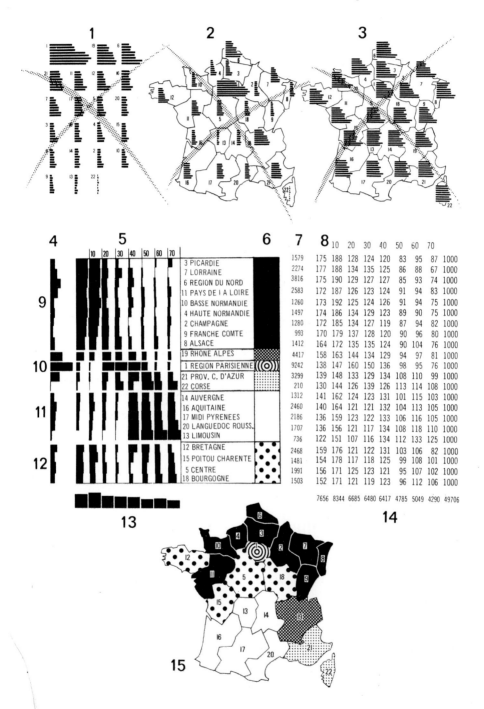

## *The image-file*  **85**

*REGIONAL POPULATION PYRAMIDS IN FRANCE.*
*IMAGE-FILE 22 × 8*

A set of population pyramids constructs an image-file. Let us consider other classic constructions: Figure **(1)** transcribes the absolute quantities; Figure **(2)** arranges them geographically; Figure **(3)** transcribes the pyramids geographically in percentages, thereby suppressing the familiar quantitative variations which are not our primary interest in statistical analysis. No serious, indisputable commentary can be derived from these displays.

*Information.* In order to see the useful information, to discover at the same time general characteristics and regional peculiarities, as well as to discover regional groupings and the demographic geography of France, it is necessary to consider the entire data set, in the form of an image-file.

*Transcription and simplification.* The absolute quantities are reduced to frequencies per 1000, enabling us to construct one card per region **(5)**. In each column black corresponds to the maximum, white to the minimum. One column indicates absolute quantities by region (7 → 4); one row indicates absolute quantities by class (14 → 13). The file can then be classed quite naturally **(5)**.

*Interpretation:* We perceive two systems: "young" **(9)** or "old" **(11)** regions; and central **(10)** or extreme classes **(12)**. Each region now belongs to a group, but its particular characteristics also appear clearly. The position of the Parisian region is indeed remarkable.
Map **(15)** shows the geographical groupings defined by the information. Note that data table **(8)** is presented here according to the classing obtained by processing the file. Each number thus acquires much more significance.

## B.2.4. — THE MATRIX-FILE

The matrix-file can be applied to a table which would normally be handled by a reorderable matrix, when one of its dimensions is too large for this type of processing. The construction is similar to that of the image-file. However, it must not exceed twenty-five to thirty columns in the fixed dimension, or implementation becomes a lengthy process.

Processing classes the file successively according to one or another characteristic.

The matrix-file is used a) to process a large number of objects across a small number of characteristics and b) to ensure that a sampling is representative.

The process of interpretation must always begin in the following way: 1st order classing based on one characteristic, 2nd order based on another, 3rd order, etc.

*FACTORY WORKERS. MATRIX-FILE 250 × 12.*

We know the geographical origin of 250 workers in a factory in central France. What are the relationships between this origin and other known characteristics (1)?

*Classing* (1). 1st order: *salary*; 2nd order: origin; 3rd order: age. We notice that *higher salaries* are paid to men (A), who are married (B), older (D), and who have more children (C) than others. Within these salaries, foreigners (G) are more numerous than French (H). Salaries have little relationship to seniority (E). *The highest salaries* (H circled) are paid to French persons from other regions. *The lowest salaries* (K) are paid to workers of local or regional origin, married, with children, relatively old and having seniority in the factory. A classing by origin clarifies these initial observations.

*Classing* (2). 1st order: *origin*; 2nd order: sex (A); 3rd order: seniority. This classing confirms that *foreigners* generally have a higher and more homogeneous salary (F) than French workers, with the exception of Iberian women (L), who are all married, young, and have children. The foreigners entered the factory recently (E) and are younger (D) than the French. The Maghrebian group (M) is the most recent. It is very homogeneous, solely male, and the group with the most children (C).

For the *French*, salaries (F) order origin as follows: 1. non-local (N), 2. township (R), 3. county (P). But local workers also have the most seniority in the factory (S). Are the qualifications of the workers from the region to blame?

*Classing* (3). 1st order: French-Foreign; 2nd order: *age*; 3rd order: seniority. Here the foreigners are clearly separated. The previous observations are confirmed. It would appear that geographical origin has more influence on salaries than age or seniority, with the exception of very young, generally single men (AB) and of senior workers from the area (2S). On the other hand, sex has only a slight influence, with the exception of the Iberian women. Is this a problem of output?

*The representativeness of a sampling.* To answer the above questions, it would be necessary to press on with the survey, to question a certain

*The matrix-file* 87

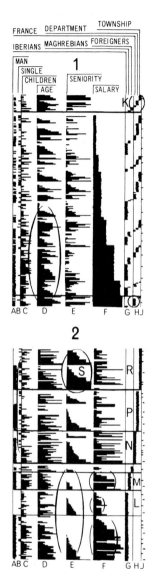

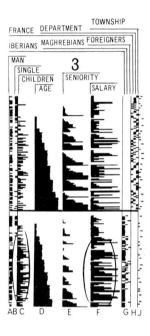

number of workers. In order to choose them, it would be sufficient to check off (J), a score of individuals regularly chosen from the 1st classing, and to do the same for subsequent classings. An additional fifteen workers in **(2)** and eleven in **(3)** provide forty-six workers, representative of all the types the data defined.

# Graphic constructions

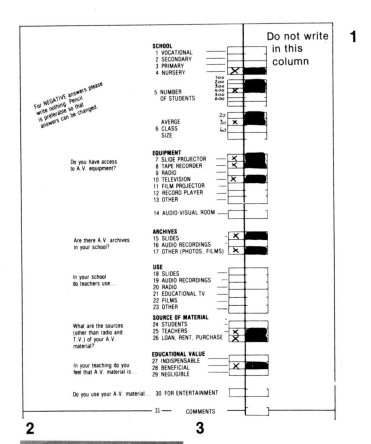

**1**

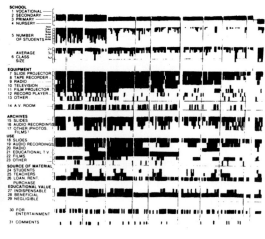

**2**

**3**

## *The matrix-file*

### A SURVEY INTO AUDIO-VISUAL METHODS IN TEACHING.*
### MATRIX-FILE 200 × 31

A correspondence survey of 200 teachers is undertaken to determine which ones should be interviewed. They must represent all types who use or do not use audio-visual equipment.

*Formulation of the questionnaire.* When the list of questions has been determined, the questionnaire is constructed as in model **(1)**. All the answer-boxes are on a single vertical line. The person being surveyed marks a cross in the first box. We later mark the second box with a felt-tip pen. In this arrangement **(1)**, the answer sheet folded in four **(2)** constructs the matrix-file **(3)** directly without any transcription.

If the questionnaire is constructed in a thoughtless way, with answer-boxes scattered all over the page, it will always have to be transcribed to be utilized. This is a source of error and a waste of time. Construction **(1)** directly creates the computer card (accessible to automatic reading) or the matrix-file, saving a lot of time and useless work. A double-entry questionnaire must always be linearized to be run through a computer, so this ought to be done when the card is initally composed. Two sheets 21 × 29.7, stuck together on the shorter side, can accommodate a line of 180 yes-no boxes.

*Utilization.* The returned answer sheets are folded and set up as a matrix-file. Classing **(3)** reveals eleven types of answer, defined by the interpretation matrix **(4)**. Then we choose two or three people per type who will be interviewed by the investigator. Thanks to **(3)** and **(4)**, the investigator positions each of his interlocutors in relation to the others. He can thereby orient the interview in terms of overall relationships which he knows and which he can then verify or disprove.

*In collaboration with R. GIMENO.

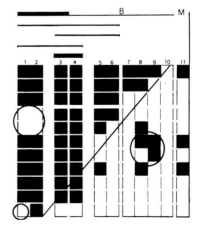

## B.2.5. THE ARRAY OF CURVES

The array of curves and the image-file are the two standard constructions for a ≠ 0 table. The curve has one advantage: it shows slopes and enables us to analyze them on all levels **(1)**. However, the quantities are represented by *y*. The rows of the matrix are therefore irregular, and the graphic technique is more complex: each curve must be constructed separately on a transparent medium so that it can be classed in various ways.

### B.2.5.1. The scale of quantities

A curve shows slopes. Consequently, a series of curves poses a problem of scale which places the designer in a position of great responsibility.

In this factory **(2)**, does production P catch up with salaries S **(2)** or exceed them **(3)** as the designer who chose the individual arithmetic scale would claim? Does production progress faster as shown by the "common" scale **(4)** or, on the contrary, do salaries, as in the common scale multiplied by ten **(5)**? Or is it true that the two now progress at the same rate as indicated by the scale of their progressions **(6)**? The numbers **(7)** show that salaries first increased by a factor of two, then by a factor of 1.5, while production increased by a factor of 1.5 in both cases. That is what is portrayed by the progression scale **(6)**.

With curves, the eye retains the intersections and the slopes, *but an intersection has no meaning when the quantities are not addable or comparable.* Figure **(3)** has no meaning. Only the slopes are meaningful. The slopes show the differences, but *the slopes are meaningless when the differences are not comparable.* The subtraction of two salaries is not comparable to the subtraction of two productions. Drawings **(2)** to **(5)** are thus visual lies whose potential consequences are obvious.

*Only the progressions, that is, the ratios between two successive numbers, are comparable in every case.* This is what is shown in **(6)**.

Consider example **(8)**. It is entitled "the strongest variations" in securities on the Paris stock exchange. Apparently the author does not know how to invest his money, since he considers that a progression from 1900 to 3300 is better than a progression from 100 to 400! The progression scale **(9)** enables us to correct the mistake, and the array of curves reveals the best investments. The classing is practically the opposite of the original!

## The array of curves

91

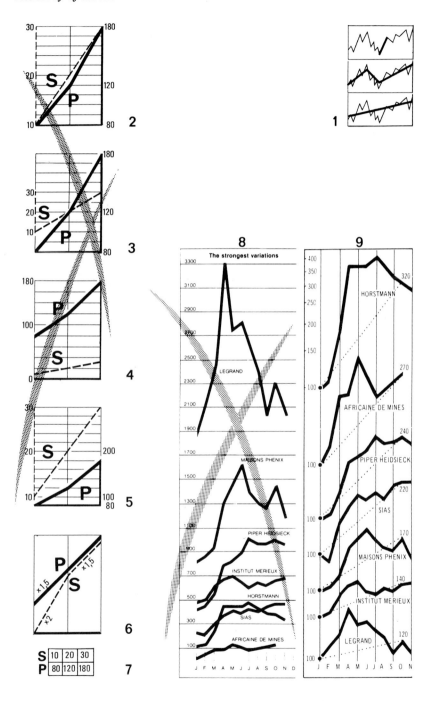

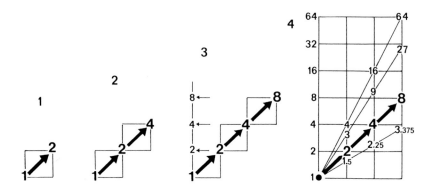

## B.2.5.2. — The progression (geometric or logarithmic) scale

The array of curves enables us to compare the evolution of many different indicators: prices, income, production, population, equipment, salaries, etc. But what can all these phenomena have in common? Only their rate of change, that is, the ratio between two successive numbers.

It is therefore sufficient to take a simple ratio in order to construct a progression scale. A production which doubles every year passes from one to two **(1)**, then from two to four **(2)**, then from four to eight, etc. If we transfer these numbers onto a straight line **(3)**, we construct a scale which expresses all the progressions **(4)**. Anyone can construct it rapidly. But to be practical, it must at least provide whole numbers. A good approximation is obtained by the method shown in **(9)**.
Remember that a progression scale does not include zero and that a displacement of magnitudes along the scale destroys the progression. However, the numbers can be multiplied by a constant quantity.

The progression scale is necessary for constructing the curves. But the reader need only know the scale of the slopes, that is, *the progression during one unit of time*. If A represents one year, scale **(5)** gives the annual progressions. Conversely, in order to calculate the progression of a given slope **(6)**, it is necessary to place the 1 on the scale at a horizontal distance A from the curve. If A represents one year, **(7)** is read as an annual progression. If A represents ten years, **(7)** is read as a decennial progression.

## The array of curves 93

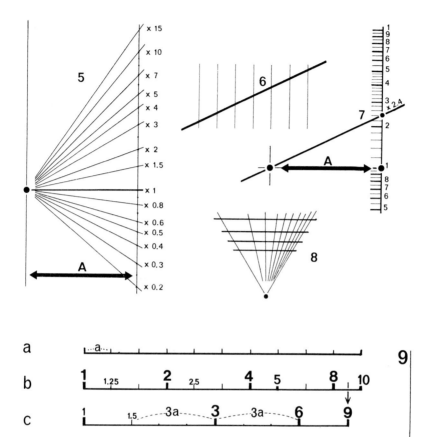

- Take ten equal spaces (a). Indicate 1, 2, 4, 8 and 1.25, 2.50, 5 and 10 as in (b).
- Mark 9 lightly, nearer 10 than 8. Mark 3 mid-way between 1 → 9 (c).
- Mark 1.5 and 6 at a distance 3a from 3.
- It only remains to mark 7, by sighting it. Record the scale on each side (d). The distance 1 → 10 is the "module" or "base" of the scale.
- "Semi-logarithmic" paper provides this scale and, in the other direction, an arithmetic scale. Choose the base by trial so that the curves are neither too flat nor too steep. Construction (**8**) permits reducing or enlarging the module.

---

*Intersections and the position of the curves on the progression scale are generally meaningless.* The curves must be classed either by types (p. 94) or by progression (**9**, p. 91). They then construct a meaningful array.

# Graphic constructions

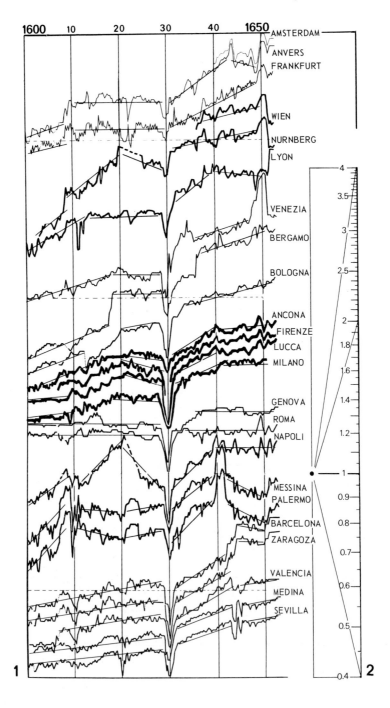

# The array of curves

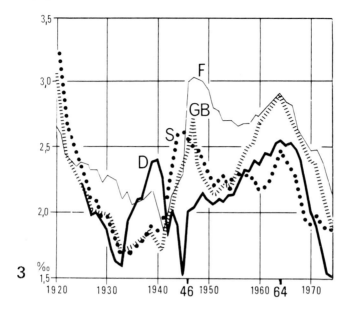

## B.2.5.3. EXAMPLES

### RATES OF EXCHANGE. ARRAY 216 × 23.

Consider, following J. Gentil da Silva and R. Romano*, the exchange rate for different currencies, at the "Besançon markets", held quarterly at Piacenza or Novi from 1600 to 1653.

A progression scale enables us to construct comparable curves directly, whatever the nature of the monetary unit.

Array (1) displays types of curves, associated with a north-south geographical classification. It reduces the twenty-three cities to seven types of financial behavior, which are more or less regional. It reduces the 216 dates to several periods, which differ, incidentally, from one type to another. Finally, it enables us to define the general tendencies, underscored by the use of linear adjustments, "smoothing lines" which we can estimate accurately by calculating their rates of decennial progression, with the aid of the scale (2).

### FERTILITY TRENDS IN EUROPE.
### SUPERIMPOSED ARITHMETIC CURVES.

To underline the great similarity in the fertility** curves of Germany (D), Great Britain (GB), France (F) and Sweden (S), they must be superimposed (3). Furthermore, the small distance between the extreme numbers (from 1.5 to three) means that there would only be a slight difference between the progression curves and the arithmetic curves, which justifies using the latter. Nevertheless, the slopes cannot be estimated accurately from this diagram.

---

*"Les foires de 'Bisenzone' de 1600 à 1650," *Annales* (1962), N° 4.

**From the Institut National d'Etudes Démographiques. Given in annual totals of women's fertility rates.

# Graphic constructions

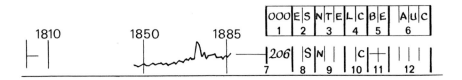

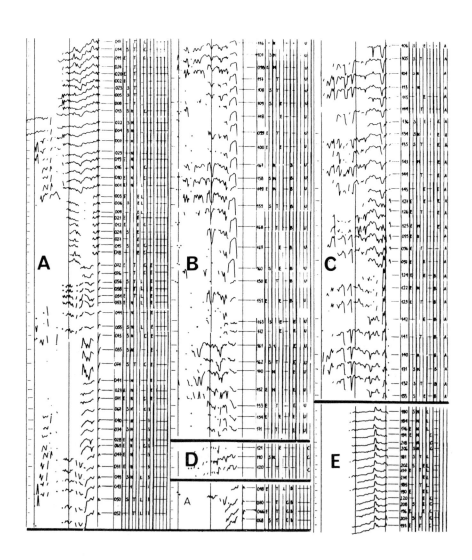

# The array of curves

*HARBOR TRAFFIC AT PARANAGUA IN THE 19th CENTURY.
ARRAY 131 × 86.*

Consider, following Cecilia Westphalen, the study of harbor traffic at Paranagua, the best statistic for indicating the development of the State of Parana in Brazil. We know a) from 1801 to 1887 the traffic at Paranagua (facing page, **A** to **D**) apportioned according to the variables (**2** to **6**) below; and b) from 1850 to 1887 the total traffic in Brazilian harbors (**E**) apportioned according to (**2** to **4**) below.

*Construction of curves.* A computer traces the progression curves. To provide good visibility, a factor not taken into account by most programs, the curves must be further intensified with a felt-tip pen and identified as in an image-file. We proceed thus:

-a) Set up the list of variables and the pattern of the code, that is: (**1**) number, (**2**) arrivals/departures, (**3**) number/tonnage/crew, (**4**) ocean/going, coastal, (**5**) Brazilian or foreign ships, (**6**) origin or destination: Argentina, Uruguay, Chile.

-b) Code each curve, always at the same distance in regard to the dates (**7**) and *always facing the last numbers of the curve*. Indicate the horizontal by carefully marking a reference line on each side of the curve. Curve 206 reads: (**8**) departures, (**9**) in number of ships, (**10**) coastal, (**11**) total ships: Brazilian and foreign, (**12**) total traffic in Brazilian harbors (a horizontal line here would mean: Paranagua total).

-c) Cut out all useless margins above and below each curve and assemble the curves on large sheets to be photographed later. Obtain 2 positive prints on transparent film, reduced to about half size.

-d) Cut out the films. The curves can now be assembled according to any given classing by similarity, or they can be compared by superimposition. Two vertical strips of two-sided adhesive enable us to fix the curves on graph paper and to remove them at will.

*Processing.* Our first assemblage enables us to see the entire set. On the facing page: (A) Paranagua total, (B) traffic with Uruguay, (C) Argentina, (D) Chile, (E) total traffic in Brazilian harbors. A great number of curves are similar. It is a matter of *discovering the meaningful variables*.

# Graphic constructions

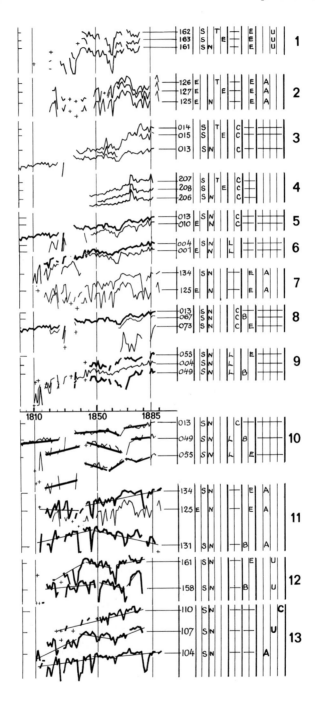

# The array of curves

*Number, tonnage, crew.* This variable has very little signification. The three curves, number-tonnage-crew are parallel. For example:
(1) Departures to Uruguay, foreign ships.
(2) Arrivals from Argentina, foreign ships.
(3) Total departures, coastal.
(4) Total for Brazilian ports, departures, coastal.
A comparison only shows the general, well-known increase in shipping tonnage. We therefore retain the most complete curve: the number of ships. It portrays the tonnage and the crew at the same time.

*Arrivals, departures.* A variable of little signification. For example:
(5) Total coastal: very clear departure-arrival parallelism.
(6) Total ocean-going: parallelism with a slight difference between 1860 and 1866.
(7) Foreign shipping, traffic with Argentina. A difference between 1870 and 1880. We only retain the departures, except for certain cases, like the Argentinian traffic.

*Coastal* (8), *ocean-going* (9). No parallelism. A significant variable.

*Brazilian, foreign shipping.* Variable with little signification for coastal shipping (8): parallelism between total coastal shipping (#013) and Brazilian (#067). Non-parallelism with foreign coastal shipping (#073). The coastal shipping is therefore mainly Brazilian, and the most complete curve (#013) can express this. A meaningful variable *for ocean-going shipping* (9). Parallel until 1850, the curves for "Brazilian" (#049) and "foreign" (#055) then reverse themselves. The total (#004) is intermediate. The ocean-going shipping is therefore divided between Brazilian and foreign ships in a meaningful way.

## UTILIZATION

*The total traffic of Paranagua* (A page 96) can thus be reduced to three meaningful curves:

(10)  #013  Departures-number-coastal-total
     #049  —     —    ocean-going-Brazilian
     #055  —     —         —       -foreign

*The geographic apportionment of the traffic* (B, C, D p. 96) is reduced in the same way to the following curves:

(11) with Argentina  #134  Departures-number-total-foreign
                           #125  Arrivals          —          —         —
                           #131  Departures    —          —    Brazilian

(12) with Uruguay     #161  Departures-number-total-foreign
                           #158         —                —          —    Brazilian

(13) For Chile, the nationality of the ships is not known. It is therefore necessary to compare:
                         #110  Departures-number-total-Chile
                         #107         —          —        —    Uruguay
                         #104         —          —        —    Argentina

These eleven curves represent the main points of the 131 initial curves. It is now up to the historians to analyze them, to compare them with the others, to discover secondary variations, to characterize the main periods defined by the curves, and to compare these periods with events *extrinsic to the data*.

## B.3. ORDERED TABLES

### B.3.1. TABLES WITH 1, 2 or 3 CHARACTERISTICS

This is the classic domain of graphics. Since it is treated at great length in numerous works, only the essential principles will be dealt with here.

*Two basic constructions.* As we saw on p. 24, tables with one to three rows offer two possible constructions depending on whether the objects A, B, C are entered along $x$ or in $z$. *Matrix constructions* (M) arrange the objects along $x$ and the characteristics along $y$. If the objects are reorderable, they must be reordered (∼∼∼). *Scatter plots* (S) are "ordered tables." The objects are placed in z, which enables us to record directly, *without permutations*. Scatter plots are used when the overall relationships are the most pertinent consideration, when the linear order of the objects is not indispensable, and when the number of objects is very large.
The transformation of (M) into (S) is schematized on page 25.

*The classification adopted here.* The table on the facing page develops the synoptic on page 29. In the center are the scatter plots (S). On either side are the matrix constructions (M). The order of the rows is reversed in relation to the synoptic. We will study in succession the transcription of reorderable objects (the left-hand part of the table), then of ordered objects (the right-hand part).

*Nomenclature*

*Matrix constructions (M: objects on x).*

1: repartition
2, 3, 4: matrices
5, 6, 7: concentration curves
12: time series
13, 14, 15: arrays of curves
Special cases:
16, 19: superimposition (subtraction meaningful)

*Scatter plots (S: objects in z).*

8 : distribution
9, 17: scatter plots with two characteristics
10: scatter plot with three characteristics
11: ordered tables (superimposition or collection of scatter plots)
Special cases:
18: superimposition of distributions
20, 21: triangular scatter plots (total per column = 100).

*Ordered tables* 101

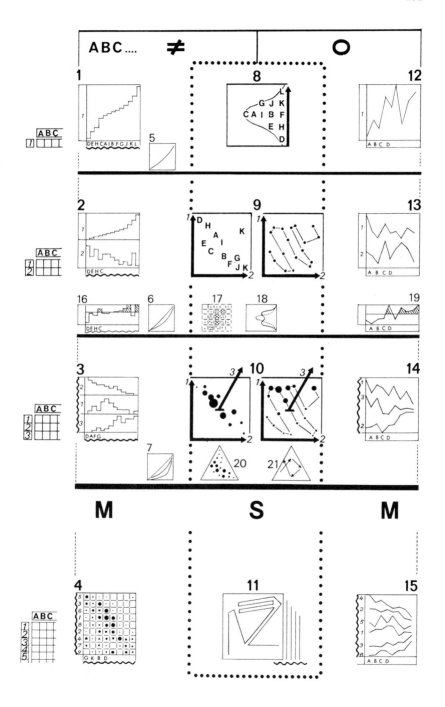

# Graphic constructions

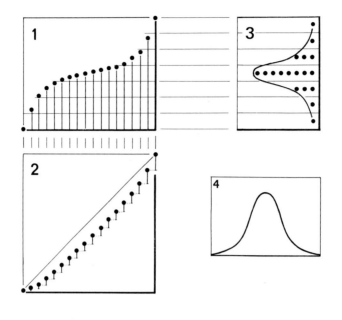

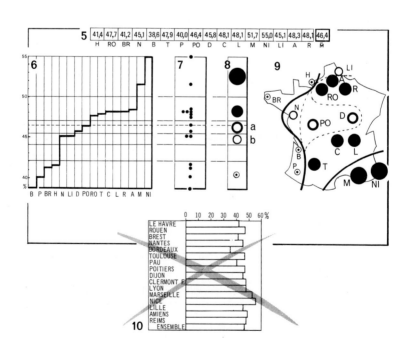

*Ordered tables* 103

B.3.1.1. REORDERABLE OBJECTS (≠). 1 CHARACTERISTIC

The goal is to discover similar or neighboring groups of objects and/or groups of quantities.

*Three constructions: repartition, concentration, distribution*

The repartition **(1)** and the concentration **(2)** place the objects along $x$ and class them by quantity.
The distribution **(3)** counts the number of objects per class of quantity. It is generally constructed as in **(4)**.
In the repartition, the groupings of objects appear in the form of "plateaus". In the distribution, they appear in the form of "humps" on the curve.

*1st example: limited A, B, C*

For sixteen French cities, take the percentage of students who answer "yes" to the question: Would you like to set up your own business? **(5)**. A, B, C are cities. We can construct a repartition **(6)** or a distribution **(7)**. If we know the topographical distribution of these cities, a map is the normal outcome of the graphic operation. *(8 → 9)*.

This example shows:
- the relationship between the repartition **(6)**, the distribution **(7)**, and the map **(9)**;
- a distribution without classes **(7)**, the most practical graphic construction for solving the problem of step series (page 197); no prior classing is necessary;
- the use of the map in the determination of types; it is in terms of geographical position that it appears useful to separate cities (a) from cities (b) **(8)**;
- the uselessness of a repartition which is not reclassed **(10)**. It betrays the designer who is unaware of the real aim of his work: discovering types and, in this case, regions.

*2nd example: very extensive A, B, C*

Take the manufacture of molded pieces ten mm wide, with a tolerance of ±0.1 mm. The mold contains eight forms. In order to check each form, forty castings are done, producing 40 × 8, 320, pieces. Thus there are 320 pieces to check, which excludes ordering them as in a repartition. *The pieces must be grouped by quantity classes* **(1)**.

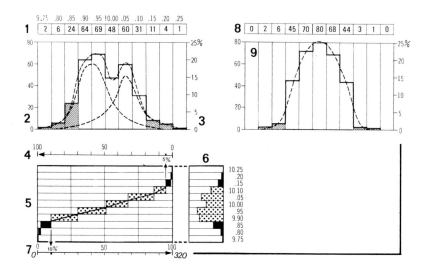

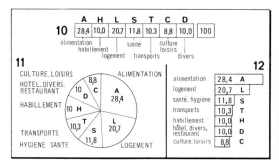

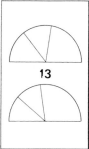

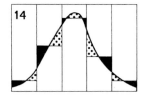
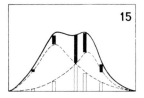

**(14)** The curve must cut out external areas which are equal to the internal areas on a histogram.

**(15)** In order to reconstitute the distributions of a bi-modal curve, the black columns are equal to the corresponding white columns.

*Ordered tables*

The pieces are measured to within 0.05 millimeters. They can therefore be enumerated in classes with intervals of 0.05 millimeters **(1)**. With a tolerance of ±0.1 mm, the enumeration showed that forty-eight pieces out of 320, that is 15%, are useless. This is an inadmissible rejection rate.

*The distribution diagram* **(2)** then shows, by its two humps (bimodal distribution), that the total number of pieces represents the addition of two populations and thus of two different operations, either by two tools or two different people, and that the only faultfree pieces actually result from irregularities in one or another operation. This greatly facilitates the correction of the molds.
When the quantity scale **(2)** is replaced by the percentage scale **(3)** we speak of a *frequency distribution.*

*The repartition diagram* **(5)** can be reconstituted from the distribution **(2)**. This is a *"cumulative curve"*. One merely reorients **(2)** as in **(6)**. Note that the "humps" in **(6)** are more visible than the plateaus in **(5)**. If we replace the scale of 0-320 pieces by a scale of 0-100 **(7)**, we can read directly that 10% of the pieces are *less than* 9.90 mm. By reversing the scale **(4)**, we can read directly that 5% of the pieces are *more than* 10.10 mm. The reconstitution of the repartition also serves to calculate the "concentration" and to compare several phenomena (p. 108).
After the correction of the forms, a new run produces the results in **(8)**; the distribution **(9)** is normal, the rejection rate of 3.75% acceptable.

*3rd example: very limited A, B, C*

Take **(10)** the division of a family budget into seven expense items* (in % of total expenditure). The very limited number of objects justifies the *circular construction* **(11)**, portraying the ratio between the parts and the whole. In **(11)** the lay-out of the nomenclature is directly accessible and avoids the costly error of colors or screens in the segments. This construction is useless in problems involving more than one row (7 p. 110), apart from exceptional cases, such as the repartition of seats in two legislative bodies **(13)**, where the vertical axis (corresponding to 50%) determines a majority.
*A repartition* **(12)** makes possible groupings, invisible in circular constructions, appear.

---

*A. Food, H. Clothing, L. Housing, S. Health, T. Transportation, C. Culture and Leisure, D. Miscellaneous.

*Graphic constructions*

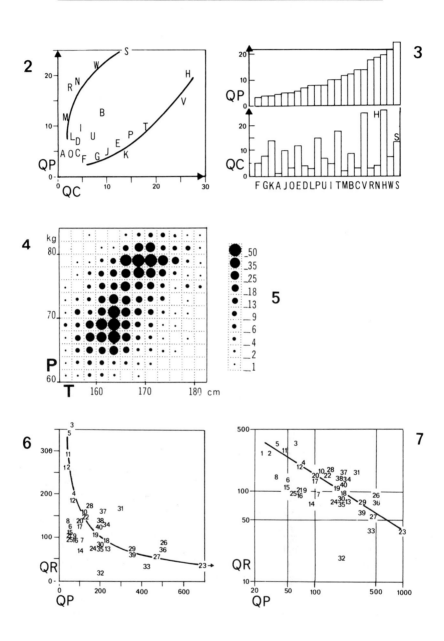

*Ordered tables* 107

B.3.1.2. REORDERABLE OBJECTS (≠) WITH 2 CHARACTERISTICS

Objective: to discover groupings of objects and/or the relationship between two characteristics. This is the privileged domain of scatterplots (correlation diagrams).

*Scatter Plots*

Take a population A, B, C, of twenty-two small towns in the Bas-Rhine* for which we know the number of inhabitants (in thousands) QP and the number of retail shops QC **(1)**.

The scatterplot **(2)** places the quantity scale QC along $x$, the quantity scale QP along $y$, and each town at the intersection of its corresponding quantities. The plot (correlation) directly displays two types of town, defined by alignments S and H. The S towns are in the suburbs of Strasbourg, where trade is not very developed. The H towns, more remote, have more highly developed trade.
For graphic comparison, the matrix construction **(3)** first orders ABC . . . according to QP and then records the quantities QC. A careful study is necessary in order to discern the two types of quantities QC, as characterized by the H and S towns.

*Scatterplot with a large number of objects.* Take a population of 620 adult males, for whom we know height and weight. We set up classes of heights at intervals of 2.5 cm (T) and weights at intervals of two kilograms (P), and we count the number of individuals defined by each pair of classes **(4)**. The quantity of objects (individuals) is transcribed by the third dimension of the image **(5)**. This highlights the presence of two ethnic groups in this population.

*Scatterplot with very extended quantities.* Take the relationship between population and income in the villages surrounding Madrid in the 16th Century. The population varies from 25-950 "households"; income per head from 17-360 units. The scatterplot can be based on an *arithmetic scale* **(6)**. It displays a "negative correlation": the greater the population, the smaller the income. This plot has an exponential form. When based on a *double logarithmic scale* **(7)**, the exponential function becomes a straight line, and the concentration of points decreases, making identification of villages and groups easier.

*From H. NONN. *L'implantation du commerce de détail.* Quoted by Sylvie RIMBERT. *Leçons de cartographie thématique.* Paris: SEDES.

# Graphic constructions

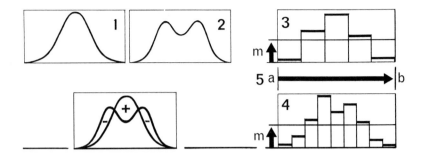

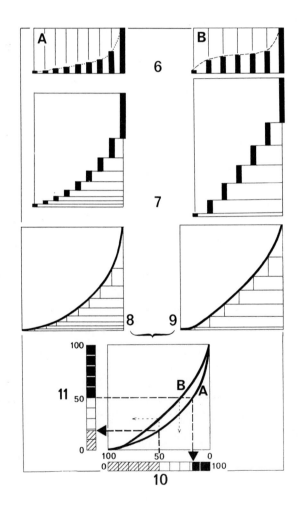

*Ordered tables* 109

*The superimposition of two distributions*
Consider a market analysis of shoe manufacturing,* which involves adjusting the *size groups* manufactured to fit the *size groups* of the population. A sample finds a dispersion between the European sizes 37 and 47, with a mean size of 42. Manufacturing will therefore vary around this mean **(1)**. The risk is great if the population is composed of, for example, two distinct ethnic types, which a slightly more precise analysis might reveal **(2)**. The result would be that a large part of the product would remain unsold ( + ), while an equivalent part of the population would be unsatisfied ( − ).

*Traps to avoid: Class intervals that are too great* **(3)** and thus hide a useful distribution **(4)**. *Irregular intervals* that deform the image. It is the columns' area which is meaningful (7 p. 191). *Different scales* between two comparable distributions. It is the total area subtended by the curve which must be equalized from one curve to the other. We first equalize the distance ab **(5)** representing the sum of the class intervals. We then equalize the two means (m).

*The superimposition of two concentration curves*

Take, for example, a comparison of income distribution over several professions or property distribution over several regions, etc. *Objects and/or classes are different*. A single question is pertinent: what are the differences among "comparable groups"? Concentration curves answer this question.

The point of departure is the "repartition" curve **(6)** that classes individuals in profession A by income. The same is done for profession B. The income quantities are cumulated **(7)**. Figures **(7)** are transformed into a square **(8)** and **(9)** that traces the concentration curves. They are superimposable.

In tracing scales from 0-100 we observe that for profession A, the first half of the individuals **(10)** possesses only 19% of the total income (for B: 29%) and that half of the income **(11)** is "concentrated" in 18% of the individuals (for B, in 30%). The concentration curve constructs a point of comparison — the first half, the first quarter, a given percentage — among data which formerly had none.

*From J.-P. SIMERAY. *Les Graphiques au service de l'entreprise*, Hommes et techniques, Paris: 1971.

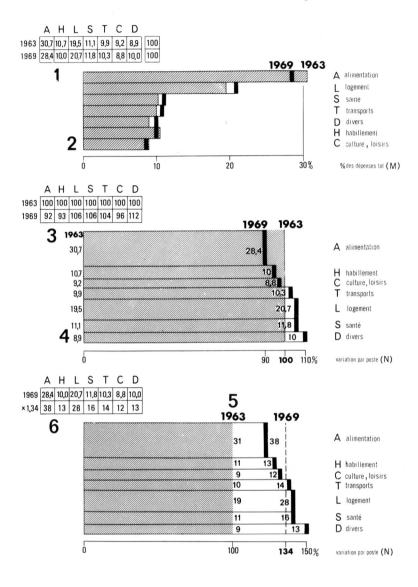
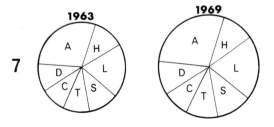

*Ordered tables* 111

*Meaningful difference: weighted diagram (Histogram)*

When the aim of the two rows of the data table is to show a meaningful arithmetic difference, the standard construction **(2)** is a superimposition of the two diagrams. However, the essential points, that is, the differences, are poorly displayed.

Take the information in **(10)** on page 104, that is, the division of the average family budget among seven expenditure items.* Here we will compare the years 1969 and 1963 **(1)**.

*The standard matrix construction* **(2)** classes the items by the 1963 quantities and superimposes the 1969 quantities. The difference is visible. We note:
- the orientation of the diagram **(2)**; it enables us to read the nomenclature easily,
- the visual means of differentiation: the transparency effect of a black line on a gray zone,
- the dominant perception: the quantitative order and obviously the perfect correlation between the two repartitions,
- but above all *the difficulty in perceiving the differences*, which is nonetheless the main goal of this comparison.

*The weighted diagram* alleviates this defect by relating the differences to a fixed base: 100 per item in 1963. This calculation **(3)** produces image **(4)**, that is, a diagram weighted on one dimension for the 1963 quantities (here vertically) and on the other dimension for the 1969 percentages compared to 100 in 1963. *The reclassing is based on the differences*, which then become highly meaningful.

Supplementary information is supplied: total consumption increased from 100 to 134 during the period. Accordingly, column 100 in **(4)** becomes column 134 for 1969 and enables us to trace column 100 for 1963 **(5)**. An elementary calculation **(6)** enables us to record on the diagram the value of each item as a percent of total consumption in 1963 (in rounded figures).

The circular constructions **(7)** are completely useless here.

---

*A. Food, L. Housing, S. Health, T. Transportation, D. Miscellaneous, H. Clothing, C. Culture and leisure—(M) % of total expenditure, (N) Variation per item.

# Graphic constructions

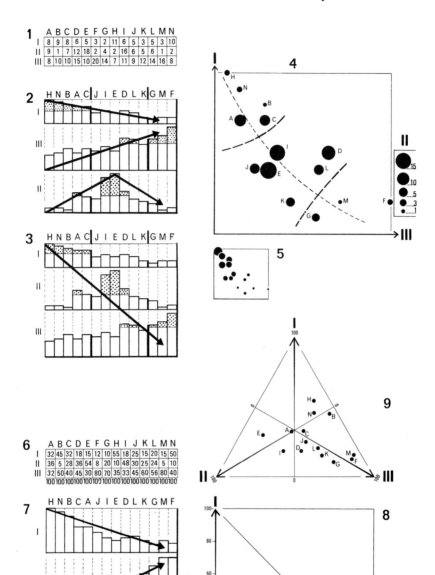

*Ordered tables* 113

B.3.1.3. REORDERABLE OBJECTS (≠) WITH THREE CHARACTERISTICS

The goals here are a) to discover groups of objects similar across the three characteristics and b) to discover, if there are grounds, two characteristics similar in relation to a third.

*When addition by column is meaningless*
This is generally the case. Examples: price/income/consumption/ production/salaries/number of workers/size of properties/ number of properties/ output . . .

*The matrix construction* is a reorderable matrix. It enables us, in example **(1)**, to discover that the objects form three groups **(2)** and that characteristics I and III are inversely related, while II constructs another system.
If the three characteristics construct an order, such as I: laborers; II: workers; III: engineers; or I: north; II: central; III: south, etc., classing **(3)** is preferable. It shows that the three groups of objects are ordered according to the order of the characteristics.

*The scatterplot* with three characteristics **(4)** places I along $y$, III along $x$ and II in $z$. It highlights the inverse relationship between I and III and enables us to characterize system II and to identify groups. A distribution such as in **(5)** would highlight a positive correlation between I and II.

*When addition by column is meaningful*

This involves a "population" divided into three classes. Example: young/adult/old; or weight of fuel/freight/passengers; or primary/ secondary/tertiary sectors . . . The total per column is adjusted to 100 **(6)**. It is then sufficient to know two rows in order to derive the third graphically. We construct a *scatterplot* as in **(8)**. *The triangular construction* **(9)** simply replaces the right angle with an angle of 60°, equalizing the three scales. Note that the three characteristics correspond visually *to the three angles* and not to the three sides of the triangle. *It is the angles which must be defined graphically.*

*Graphic constructions*

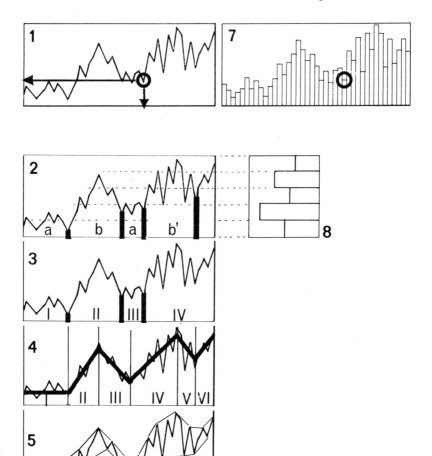

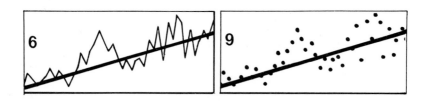

## B.3.1.4. Ordered objects (0) with 1 characteristic

The aim is to discover groups of objects (periods) and/or groups of quantities.
*The time series ("Chronogram")*
When objects A, B, C are ordered, by date, for example, the standard construction is the time series **(1)** or chronogram. It furnishes all three information levels:

*The elementary level.* On a given date, how much? **(1)** or conversely: for a given quantity, when? At this reading level drawing **(7)** is more precise.

*The intermediate level.* Regrouping quantities into categories of quantities a, b . . . **(2)** enables us to define types a, b . . . in the series of dates and sub-types b, b', based on the precise structure of the curve.

Regrouping objects, that is, defining successive periods, can be based either on absolute quantities **(3)** or on their ratios, that is, slopes **(4)**.
The determination of mean (or adjusted) slopes or "linear smoothing" **(4)** can be done by calculation (moving averages, sliding scales) or visually. It has been shown that for average complexities visual smoothing is often superior to that obtained by calculation, which always depends on the smoothing algorithm chosen. The straddling or "artillery" method demonstrates remarkable precision here. Trace out one curve slightly too high, another slightly too low. The central curve is the good one.
The drawing of an "envelope" **(5)** can also reveal general tendencies as well as exceptional or erroneous points.

*The overall level of the entire set* defines the general relationship between the two components, for example, "prices increase over time." This is rectilinear smoothing **(6)**, easier to trace on a cluster of points **(9)** than on a curve **(6)**.

*The distribution* **(8)** enables us to make the categories of quantities appear in a more precise way.

**116**                                              *Graphic constructions*

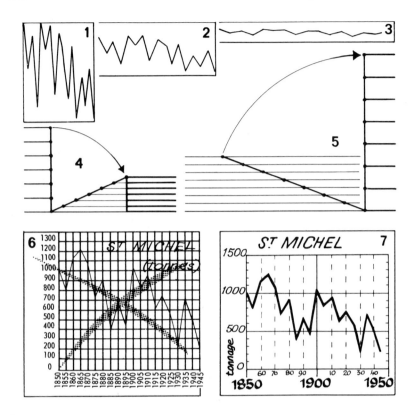

*Problems of scale in "time series"*

Avoid **(1)** and **(3)**. The elementary angles must be of average openness **(2)**. Figures **(4)** and **(5)** illustrate a graphic procedure which enables us to reduce or enlarge any scale.

When periods are characterized by slopes, remember that only the progression (or logarithmic) scale allows us to compare slopes (p. 90).

Different scales, such as the Gaussian (p. 228), level off a known general tendency in order to illustrate only variations in relation to this tendency.

*Problems of visibility* (p. 228)

Avoid a drawing such as **(6)** where the grid is more visible than the curve. Identification of $x$ and $y$ must be immediate; record this identification directly along $x$ and $y$. Three dates are immediately legible in **(7)** and

*Ordered tables*

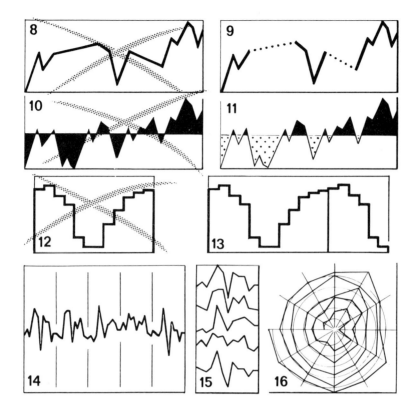

enable us to proceed to a precise identification. The twenty dates in **(6)** are useless and cumbersome. The same holds true for the specification of tonnage along $y$.

Unknown data must be clearly identified as such **(9)**, or else the drawing is falsified **(8)**. Avoid **(10)**; drawing **(11)** is much more efficacious, particularly for comparing numerous curves.

*A cycle* can only be seen clearly if its two periods are drawn completely **(13)**. Annual or weekly cycles, etc., are groupings of objects. When the pertinent question concerns the general trend, construction **(14)** is correct. When it is a matter of comparing the development of cycles, **(15)** is better. But we must not forget that the construction of deviations from the mean curve or mean cycle is always more efficacious. Construction **(16)** is much more difficult to read and interpret than **(15)**.

**118** *Graphic constructions*

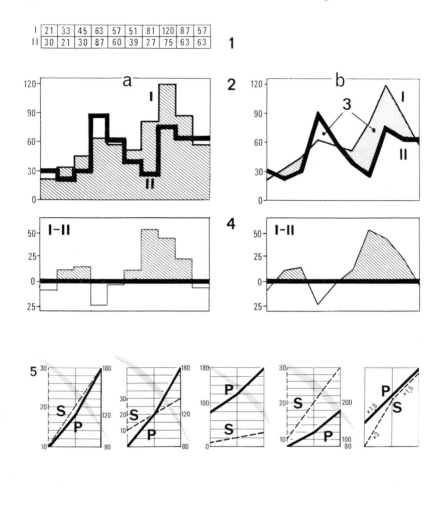

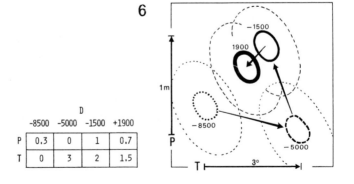

*Ordered tables* 119

## B.3.1.5. Ordered objects (O) with 2 characteristics

Object: to discover "periods" defined by two characteristics.

*Time series*
*When subtraction is meaningful: superimposition and arithmetic scale.*
When subtraction or addition is meaningful, the construction must superimpose the two curves **(2)** and add either the curve of the difference **(4)** or the curve of the total.
Example **(1)**: comparison by year of the number of houses built (I) and the number of houses demolished (II). The difference is the number of new houses available. The balance of payments is the difference between imports and exports . . .
Formula (a) more accurately delineates the area. Formula (b) is more continuous and more legible. In **(3)** the gray-white contrast is more legible than the contrast between the two different screens.

*When subtraction is not meaningful: juxtaposition and logarithmic scale.* When the counted units are different or disproportionate, only the slopes are comparable, which requires use of a logarithmic scale. Remember **(5)**, the example from page 90, comparing production P with salaries S.

*Scatterplots* **(6)**.

1st example: the evolution of "ombrothermic" area.* Each vegetal ecosystem reflects a certain relationship between measured precipitation (P) and temperature (T). Successive strata of pollen at a given site define successive vegetal ecosystems. That is, they define the conditions P and T for each layer of pollen, and that defines each period (D). We can thus reconstitute the evolution of the climate at this site.
The numbers define the difference in relation to the smallest number, designated zero. The variations amount to one meter for P and three degrees for T. The evolution of the climate is indicated by the line which joins the successive periods.

*From P. REY. *Phytocinétique biogéographique.* Paris: CNRS, 1960.

**120**  *Graphic constructions*

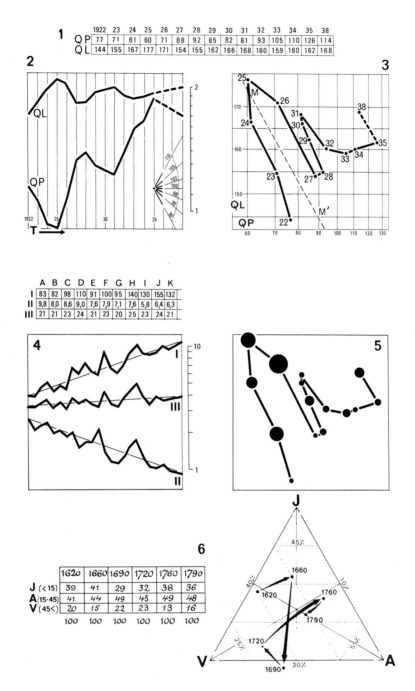

*Ordered tables* 121

*2nd example: the elasticity relationship.* The scatterplot has a particular utility when the two characteristics have a probability of reacting directly to each other, with or without a time gap. Such a relationship is considered to be elastic. The scatterplot makes it possible to display this elasticity and to define its constancy. Take **(1)**, the real price (related to the price index) of postal rates (QP) and the number of letters posted (QL) in millions.* The *superimposition of two series* **(2)** only manifests the contrast between the two curves.**

*The scatterplot* may also place the prices along $x$ and quantities along $y$ **(3)**. The two scales are logarithmic, since it is now important to define the slope. But here the scales do not need to be constructed on the same base (module). The years are points. *The succession of years is defined by a line.* We see that the increase in quantity corresponding to the reduction of price resembles the reduction of quantity when prices increase. This constant is defined by the direction $MM^1$ whose characteristics can be expressed as a ratio of $x/y$.

### B.3.1.6. Ordered objects (0) with 3 characteristics

The goal: 1) to discover "periods" defined by three characteristics; and 2) to discover, where possible, two characteristics that are proximate in relation to a third.

*The array of curves* is in the form of a logarithmic scale and compares the slopes **(4)**. Two similar curves must be brought together (1 and III for example). Linear smoothing enables us to characterize the periods while taking the three curves into account. The superimposition of arithmetic time series is justified in the case of very comparable curves (page 95).

*The rectangular scatterplot* **(5)** superimposes a third characteristic in $z$ on an "elasticity" diagram.

*The triangular construction* **(6)** conforms to the same conditions as on page 113. It enables us to characterize different periods according to, for example, the repartition of the population into young (J), adult (A), and old (V).

---

*From MORICE and CHARTIER, quoted in PEPE: *Présentation des statistiques.* Paris: Dunod, 1959.
**Note in **(2)** the drawing of the logarithmic scale. Each slope is defined by its annual coefficient of variation. The module (base) only ranges from 1 to 2.

# Graphic constructions

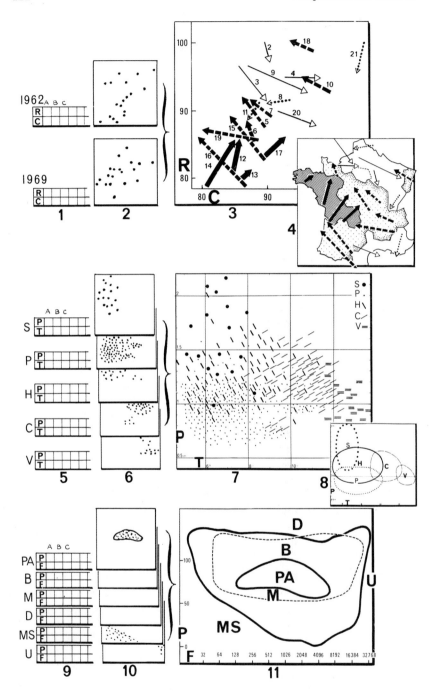

*Ordered tables* 123

*B.3.2. SUPERIMPOSITIONS AND COLLECTIONS OF TABLES*

Data tables with more than three rows are generally transcribed by *matrix constructions*, as discussed in section **B2** of this book. But it can be useful to superimpose several scatterplots, which will construct an *ordered table*, or to construct a *collection of ordered tables,* that can be classed. A last possibility is a triangular scatterplot which can convey a fourth characteristic in $z$.

### B.3.2.1. Superimpositions of tables: the "ordered table"

When objects have either qualitative (yes/no) characteristics or a pair of ordered or quantitative characteristics, it is often useful to construct a table placing the two ordered characteristics along $x$ and $y$ respectively. When each qualitative characteristic is constructed in a separate table, the set constitutes a *collection of tables* that can be compared and classed in different ways (pp. 125 and 126).

When the distribution is sufficiently simple, the tables can be superimposed, constructing an *ordered table* **(3)**. Like any superimposition of images, the ordered table raises the graphic problem of the *visual selection of the different sets* (pp. 169 and 213).

Take a comparison by regions 1, 2, 3 . . . of the relationship between income (R) and consumption (C) in 1962 and 1969 **(1)**. The superimposition of two scatterplots **(2)** enables us to construct image **(3)**, which provides the legend for map **(4)**.

Take a comparison of the characteristics of temperature (T) and measured precipitation (P) for different trees **(5)**.* Each tree constructs a scatterplot **(6)**, which can lead either to an exhaustive **(7)** or to a simplified **(8)** superimposition. The "audible area map"** portrays different sounds **(9)** — speech (PA), industrial noise (B), music (M), painful sound (D), micro-sounds (MS), ultra-sound (U) — characterized by their frequency (F) and their volume (P). The superimposition of the area of each sound constructs **(11)**.

---

*From P. REY, cited earlier. S. Fir, P. Pine, H. Beech, C. Oak, V. Holm-Oak.
**From A. MOLES. *Théorie de l'information et perception esthétique.* Paris: Flammarion, 1958.

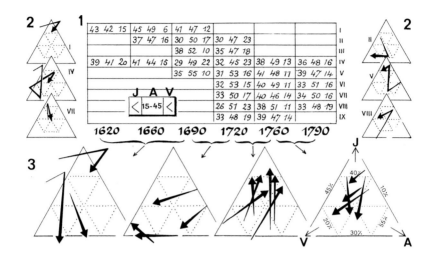

### B.3.2.2. Collections of tables.*

#### DEMOGRAPHIC MOVEMENTS IN THE 17th CENTURY**

Take the percentage of young (J), less than fifteen years old, adults (A) and old (V), more than forty-five years old, in villages I, II, III . . . , from 1620 to 1790 **(1)**.

We can construct one triangular scatterplot per village **(2)**, making characteristic periods appear **(6** p. 120). If these periods are common to several villages, the construction of one scatterplot per period **(3)**, superimposing all the villages, shows the applicability of the proposed periods.

#### PORTRAITS OF ESKIMO HUNTERS***

Take ninety-two eskimo hunters. For each, and for specific years, we know the hunting ground. Can we define types of hunters from this information? A table, showing the years along $x$ and the hunting grounds ordered from north to south along $y$, enables us to construct a "portrait" of each hunter.

The ninety-two portraits can be classed in different ways. On the facing page they are classed from left to right by age and from top to bottom according to geographical preference: northern at the top, southern at the bottom, those with no preference in the center.

Within each age group two columns, where applicable, specify two different tendencies: hunting more and more to the north or more and more to the south. An arrow underscores these tendencies.

In conclusion we may note that they all operated in the neighborhood of KRANAQ during the 1960's, except for a few who are marked by a circle.

*See also p. 263.
**From R. BAEHREL. *La Basse Provence rurale au 17e siècle.* Paris: SEVPEN, 1961.
***From JEAN MALAURIE. *Etude des Esquimaux de Thulé.*

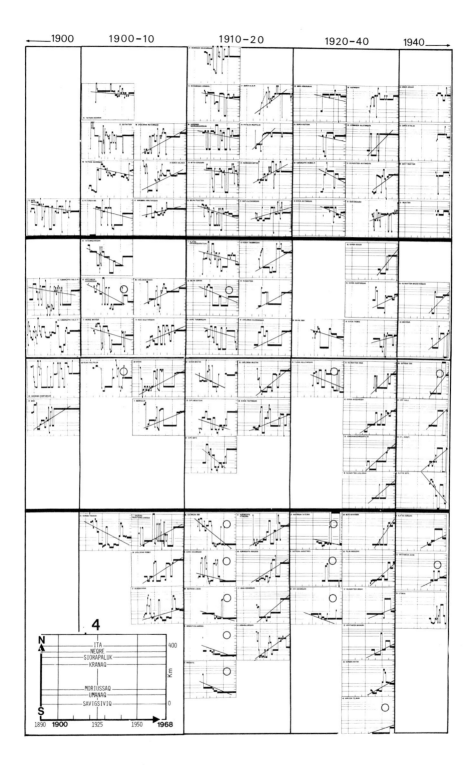

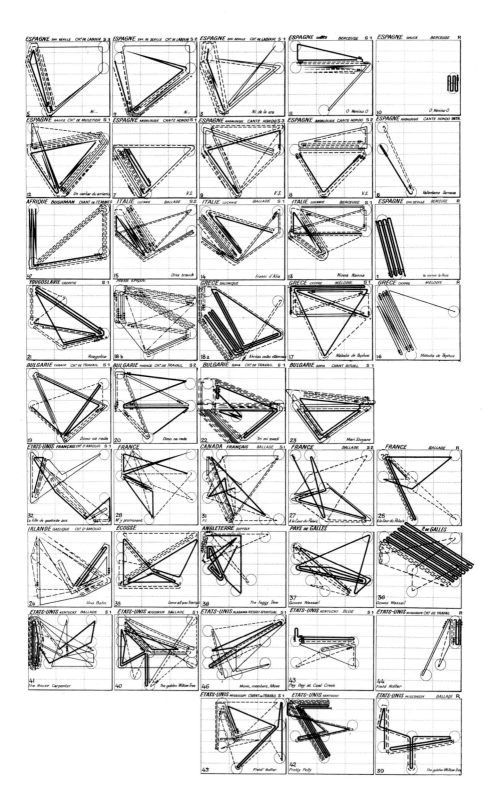

# Ordered tables

 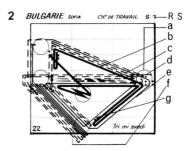

R. Refrain    S. Stanza
a-circle indicating the sounds used
d-prolonged sound
b,c,e,f,-1st, 2nd, 3rd, 4th phrases
g-same sound repeated twice in a row

## FOLK SONGS*

Consider a collection of folk song recordings. This collection cannot be studied profitably unless it is transcribed into a sign system that will facilitate comparison. The graphic image is one such system.

Only the vowels are retained here. The set of vowels can be inscribed on a "vocalic map" (**1**), which organizes them along $y$ by position of the tongue and along $x$ by the "anterior (I)-posterior (III)" variation (throat sounds mouth sounds) and the rounded (A) or non-rounded (N) form of the lips.

This ordered table enables us to draw each song couplet by connecting successive vowels (**2**). The phrases are distinguished by different line patterns. Repetitions are noted. The parallelism of lines joining the same sounds is the key to the legibility of these images.

The collection allows for different experimental classings. The songs are classed on the facing page from top to bottom by increasing complexity. The refrains are in the last column on the right. We can observe that the Romance languages are not opposed to the others, but that the Mediterranean languages form a group of much simpler images in contrast with the complex images of English, French, Gaelic . . . . We may note the simplicity of all the refrains (R) and the tendency of the lullabies toward a horizontal line, also found, incidentally, in melodies, religious chants and the blues! Many other observations are possible, and we come to regret the lack of more examples for each language (do all Spanish songs end, as here, in a triangle?), and for other languages, such as German, Russian, Arabic, and Asiatic languages, not represented here.

*From A. LOMAX and E. TRAGER. "Phonotactique du chant populaire," *L'Homme,* Paris, 1954.

**128** *Graphic constructions*

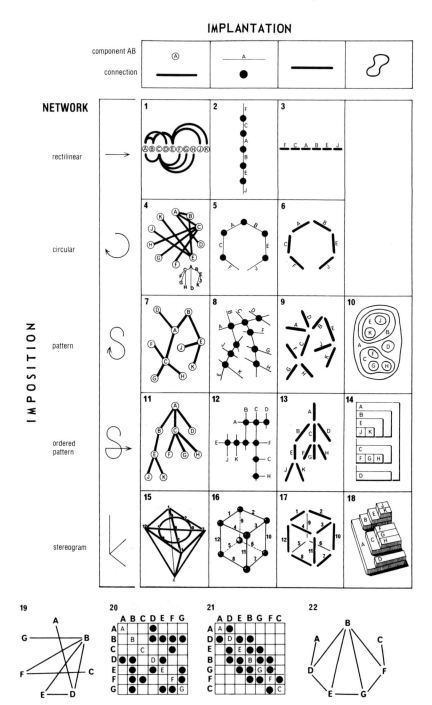

## B.4. REORDERABLE NETWORKS

### B.4.1. Network constructions

The relationships established between two sets of elements constitute a *diagram*. The relationships established between the elements of a *single set* constitute a *network* (p. 192).
Take the information: A is the father of B, C, D; C is the father of F, G, H; B is the father of E . . . This "genealogical tree" delineates the network of kinship relations among individuals A, B, C, . . . **(1)**.

*Means of representation.* The individuals can be represented by points and the connections by lines (1st column on the facing page) or conversely (the 2nd column). In certain cases lines alone (3rd column) or areas (4th column) can represent both at the same time.
Furthermore, these "implantations" can be arranged in a linear **(1)** or circular **(4)** manner; in a pattern **(7)** or an ordered pattern **(11)**; or they can suggest a volume **(15)**. The combination of these "impositions" with the three implantations defines the types of network construction as provided on the facing page.

*Why draw a network?* As with any graphic, networks are used in order to discover pertinent groups or to inform others of the groups and structures discovered. It is a good means of displaying structures. However, it ceases to be a means of discovery when the elements are numerous. The figure rapidly becomes complex, illegible and untransformable.

*Simplification of a network.* Take seven individuals positioned around a table. They establish verbal relationships **(19)**. These relationships would have been easier in arrangement **(22)**. In such cases, visual simplification is possible, as we will discuss later. However, with thirty individuals or more, correct simplification is impossible, without resorting to mathematics and graphs.

*Matrix constructions.*

The double-entry table may be used to construct a network, if A, B, C . . . are recorded twice: in rows and in columns. Relationships are represented by points, and the matrix is permutable. Thus **(20)** represents **(19)**. Permutations lead to **(21)**, which produces **(22)**, provided **(21)** is read from the central element B, that is, from the element which has the most connections. This procedure remains complex and is currently under study.

**130** *Graphic constructions*

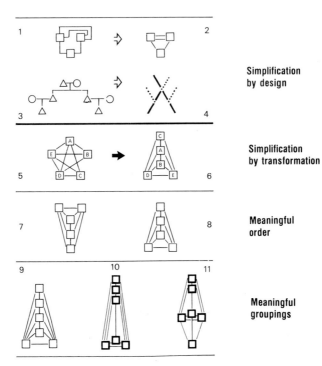

Simplification by design

Simplification by transformation

Meaningful order

Meaningful groupings

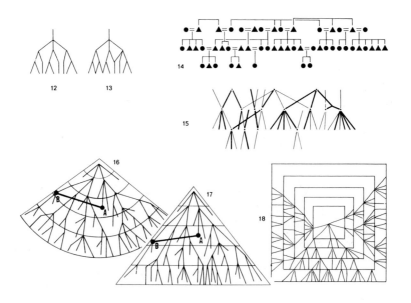

*Reorderable networks* 131

## B.4.2. Simplification and transformation of a network

*Simplification by design.* Very often a network, "flow-chart" or "tree" is unclear simply through the fault of the designer. Remember that any angle destroys the unity of a line. A network is often complex in itself, so it is necessary to eliminate all needless complexity and to replace a figure like **(1)** by **(2)**. In the same way, a "tree" representing individuals by lines **(4)** simplifies the reading considerably.

*Simplification by transformation* is the main problem with networks. By arranging the elements in a circle **(5)**, any connection can be transcribed by a straight line. This is the construction which enables us to pose the problem graphically. How do we subsequently develop from **(5)** to **(6)**? Lacking a simple and general method of calculation, we must progress through successive trials, attempting to reduce the number of meaningless intersections.

*Meaningful order.* Figure **(7)** does not have the same immediate meaning as **(8)**. The plane is naturally ordered from top to bottom by gravity. The left/right order is much weaker. The final drawing must take this natural order into consideration and attempt to make the vertical axis correspond to a meaningful order and the horizontal axis to an equality.

*Meaningful groupings.* Simplification and order are not sufficient. Figures **(9)**, **(10)** and **(11)** are not similar, and it is much easier to understand and remember three groups **(11)** or two groups **(10)** than six elements **(9)**. This is precisely the problem in information processing. The "good" drawing of a network is indeed a form of data processing which must portray the meaningful groups.

*"Trees"*

Figures **(12)** and **(13)** are networks. When there is only one way to go from one element (node) to another, the network is a "tree" **(12)**. Complex figures such as **(14)** finally become readable **(15)** when points are replaced by lines. Genealogical trees ordered by generation may adopt the sector construction **(16)**, but the triangle **(17)** is preferable (A is clearly anterior to B). Very numerous populations lead to the square **(18)**. The superimposition of male and female networks is always complex. Construct two separate networks, to be superimposed only when printed. Thus we have three figures, of which two are legible.

# Graphic constructions

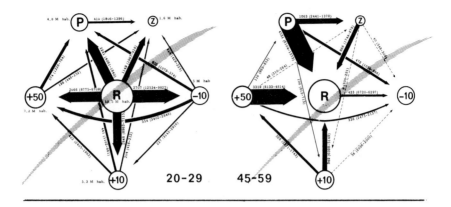

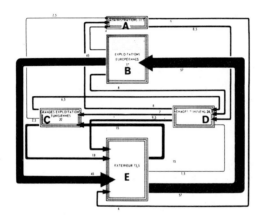

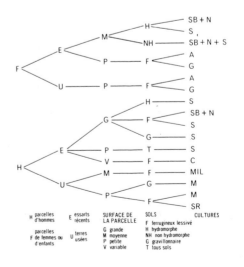

## B.4.3 EXAMPLES OF TRANSFORMATIONS

RURAL EXODUS*: migration of French voters twenty-one to twenty-nine years old and forty-five to fifty-nine years old in 1953. *The elements* are six types of community: Paris (P), Paris suburbs (Z), communities with more than 50,000 inhabitants (+ 50), more than 10,000 (+ 10), fewer than 10,000 (− 10), and rural communities (R). They are weighted by the total population per category.

The *connections* are oriented by net movement and weighted. A study of the figures places the example of the rural communities before us (R). Placed at the center of the pattern, they construct two starkly contrasting images. But these rural communities thus overwhelm all the other groups, and the careful reader is not satisfied. The types of community are ordered. What happens to this order in relation to total migration? The answer requires a transformation (following page).

FINANCIAL EXCHANGE IN A MARKET ECONOMY: Five *economic groups* — administration (A), European farming (B), Tunisian farming (C), Tunisian households (D) and outside interests (E) — enable us to analyze *financial exchanges*, by orientation and weighting. The pattern on the facing page is arranged so that the inputs are situated on the same side for each group. The figure is an illegible inventory, though the network itself is very simple!

SHOULD WE RETAIN THE "TREE" FORM?

A study of land lots in an African region** takes different characteristics into consideration: men's lots (H) or women's lots (F); newly cleared (E) or farmed-out land (U); types of soil: ferruginous (F), hydromorphic (H) or not (NH), gravelly (G) or all types (T); and finally types of crops: white sorghum (SB), nib (N), sesame (S), peanuts (A), cotton (C), millet (MIL), red sorghum (SR), maize (M). Interpreting a matrix-file enables us to construct the tree on the facing page. But what are the characteristics of the women's lots? The characteristics of the sesame crop? The soil for each crop? A "tree" must always be read point by point. Its answers are never in visual terms.

*From P. VIEILLE and P. CLEMENT. *Etudes de comptabilité nationale.* Paris: Ministère des Finances, 1960.
**From Michel BENOIT, *Espaces agraires Mossi en Pays Bwa (Haute-Volta),* t. l. ORSTROM, 1973.

*Rural exodus*

Ordering the types of community in terms of size (not total population) makes it possible to distinguish "ascending" and "descending" movements. The perception of types becomes much easier. They no longer have to be read one by one. The contrast between the two age groups is obvious. Furthermore, each type assumes its place in the general movement, and we notice that for young people, all types of community, not just rural communities, have an ascending net movement, with the obvious exception of Paris.

*Financial exchange*

The first operation consists of eliminating all meaningless angles. This is the aim of **(2)**. The transformations in **(3)** and **(4)** eliminate all meaningless intersections. Figure **(4)**, the simplest, is not the easiest to understand. We must separate out the groups. Figures **(5)** and **(6)** contrast the farming operations with the general economic groups and enable us to compare the farming operations.

*The "tree" form*

The tree destroys the unity of the groups. For example, it does not permit recording the soil for each crop. Yet, a "tree" may be manipulated, and we must take care to make the output (here along $y$) as homogeneous as the input along $x$. The network on the facing page provides answers to the three questions posed earlier.

# Reorderable networks

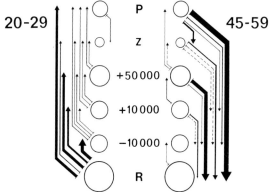

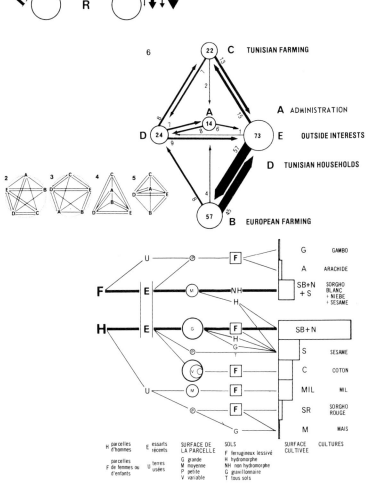

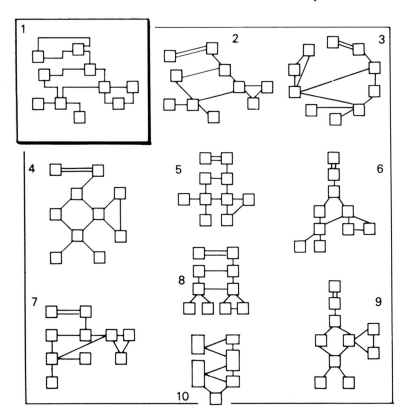

## B.4.4. The objective of "flow-charts" and "organigrams"

To construct a flow chart is to process information. Consider the programming operation depicted in (1). We must begin by eliminating meaningless features. We obtain (2). What is the longest line? Figure (3) displays it. It is now up to the specialist to choose among a quincunx (4), three groups (7), two groups (5 and 8), or a linear system (6 and 9). Examining this problem the specialist discovers that (8) is preferable to all the others and that it permits new simplifications. Figure (10) is the one finally retained. Ordering and grouping are the province of information-processing.

*Order and disorder* are very strong visual perceptions. Their application enables us to characterize (on the facing page) two well-known French "constitutions" at a pedagogical level.

# Reorderable networks

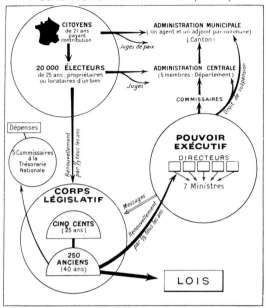

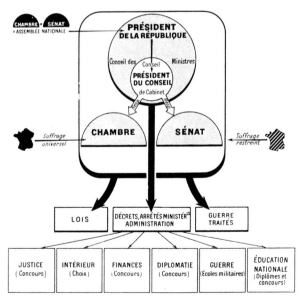

**Graphic constructions**

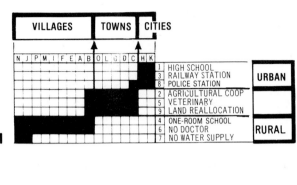
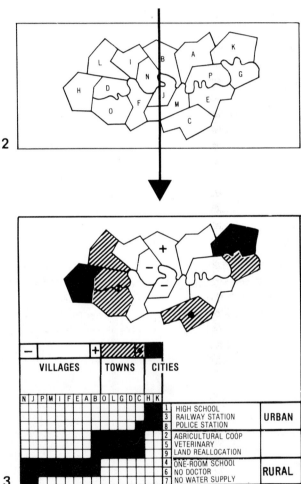

# B.5. ORDERED NETWORKS: TOPOGRAPHY AND CARTOGRAPHY

A topography is a representation of a physical object. More precisely, it traces, at a given scale, the natural arrangement of certain elements of an object. *A topography is an ordered network.* The type of order is determined by the object, which may be a geographical area, the heavens, a human being, a piece of furniture, a machine, etc.

## B.5.1. INFORMATION PROVIDED BY A MAP

A geographical map is a representation of the arrangement of elements on the surface of the earth. The order defined by the terrestrial surface confers two exceptional attributes to the map.
- The map supplies *intrinsic information as to topographical proximity*, which only it can transcribe completely.
- The map constructs a constant and universal reference shape, constituting the most powerful means of introducing into the problem *the extrinsic information necessary for interpretation and decision-making.*

A regional planner might suggest, if not insist, that the map is now useless in comparison with modern means of information-processing. Let us examine the information that this expert neglects, through simple example. Information-processing enables us to discover groups of similar townships (p. 32) to which we could apply similar planning decisions. But are these townships **(1)** homogeneous? The *topographical proximity* information **(2)** reveals **(3)** that, all things being equal, only one of the groups, N . . . B, is homogeneous. For the two other groups, the map discloses that "all things are not equal" and that it is necessary to modulate the decisions.

Moreover, the map also yields *extrinsic information*, from two perspectives. It constructs a *constant reference shape* such that all maps of the same region are immediately comparable. Everything that we have unconsciously memorized in the form of a map is automatically introduced into the problem. Furthermore, it adds *general geographical knowledge.* Townships along rivers are not homogeneous with the townships on plateaus; each group has particular characteristics, already known and defined. The same is true for townships on the plains, in the mountains or suburbs. The climatic, geological, economic and other conditions imply general constraints already identified by numerous studies.

These two additions — intrinsic relationships of topographic proximity and the introduction of extrinsic information — are indispensable for interpretation and decision-making and are the two specific properties of cartography.

*The main problem in cartography* is the counterpart of these very properties. In a diagram, the geographic component AB . . . only utilizes a single dimension of the plane. The other dimension remains available for transcribing $n$ characteristics. In a map, the component AB... constructs a network that utilizes the entire plane, in fact, accounting for the map's effectiveness. But the $y$ dimension of the plane is no longer available for the representation of the characteristics, so we must choose between two solutions:
- *either construct one map per characteristic.* In this case the map answers two types of question: Where is a given characteristic? What is there at a given place?
- *or superimpose all the characteristics on the same map.* But then the question: where is a given characteristic? no longer has a visual answer. Should this question indeed have an answer? This is the basic problem in cartography with $n$ characteristics, that is, "thematic" or more precisely, "polythematic" cartography.

Thus we will examine successively:
- How to construct and transcribe the topographic network, that is, the base map.
- How to transcribe an ordered characteristic in $z$.
- How to transcribe several different characteristics.

## B.5.2. THE BASE MAP

In order to map the number of inhabitants per hut in an African village, the ethnologist must first draw a map of the huts, that is, determine their respective positions along $x$ and $y$ on a sheet of paper. This is a *topographic* operation, done on a drawing board.

It remains to transcribe for each of the huts, that is, for each position $xy$ on the sheet of paper, the quantity of inhabitants, $z$. This transcription is a *thematic* operation. The quantity $z$, here, represents inhabitants. On another map it might represent income. On others it might represent the age of the hut, its size, its height above river level, the number of children, the father's job . . . The "themes" are innumerable, but the $xy$ topography remains constant.

### B.5.2.1. The construction of a topographical network

Geographical topography is applicable to two problems:
- determining distances $x$ and $y$ on the sheet of paper among appropriate reference elements; this involves triangulation and planimetry;
- finding the most *workable* solution to the unsolvable problem of imposing a sphere onto a plane. This is the problem of cartographic projections.*

This is not the place to discuss such questions or their highly technical modern solutions. Geodesy, topography, and geometry do serve to provide the "base map." But the problem remains poorly solved and is sufficiently difficult for specialized organizations to devote themselves entirely to it, *without confusing topography with thematics.*

*Topography itself does not pose a problem of graphic notation.* Its only task is to supply points and lines. The representation of relief, for example, is not a topographical question. Though the topographer may be the best equipped to define the altitude of a point, his need to exclude it from the $xy$ calculation for this point leaves him no better equipped than the thematician to choose among the multiple means of representing the variations of this quantity in $z$. Accordingly, the inviolate contour curve

---

*S.G., p. 288. Here and henceforth SG refers to J. BERTIN. *Sémiologie graphique.* Paris and La Haye: Mouton, Gauthier-Villars, 1973.

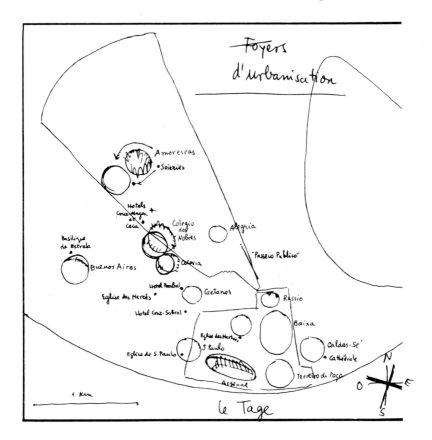

allows very significant differences in relief to escape, sometimes overlooking them because they are situated between two contour curves, because their base is not horizontal, or because the generalization was done without sufficient knowledge of the geographical characteristics.

### B.5.2.2. The base map: practical problems

The base map is supplied by the topographer, but the problem of its geographical use remains complex. The thematician encounters the following difficulties: the sought-after information and the statistical boundaries are given on too large a scale; the available maps are overloaded; they are often in color, which makes copying impractical. Consequently, the thematician is often obliged to re-draw the base map.

## Ordered networks: topography and cartography

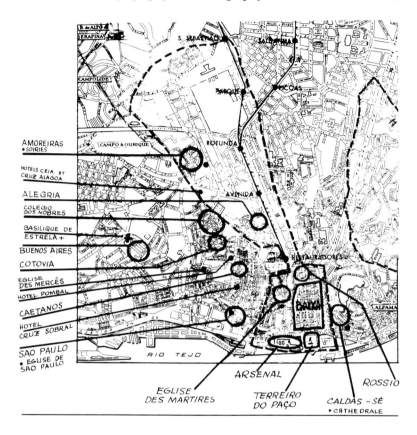

*The recording of new topographical information.* Precise positions, sites, lines and areas unique to each problem must be recorded *on the map itself* rather than on an overlay. The nomenclature and definitions should be written in the margin, always in CAPITAL LETTERS. If need be, the margin can be enlarged by sticking a strip of paper onto it. The written keys are connected to corresponding sites by lines **(2)**.

*The base map should only be drawn once,* and in such a way that it can be reproduced for worksheets and also for final publication. The means of copying should be determined beforehand.

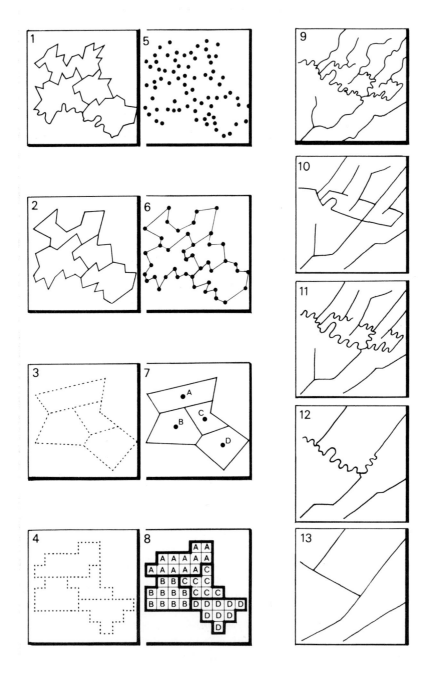

*Ordered networks: topography and cartography* **145**

*The base map can be highly simplified.* The level of simplification or "cartographic generalization" depends on the type of information sought.

*1st example: administrative boundaries.* Take the topographical information **(1)**. Should it be drawn as in **(1), (2),** or **(3)**? The answer is a question: What groupings are sought? Those in **(5) (6)** or **(7)**? If the pertinent question is: how are the areas grouped? A-BCD or AB-CD or AC-BD, etc., drawing **(3)** provides the necessary and sufficient relationships in the most legible manner and is therefore best. In on-line computer cartography, the drawing is reduced to **(4)**, which corresponds to **(8)**, a "grid" map or "mesh" map.

*2nd example: a hydrographic network.* Take the topographical information **(9)**. It can be simplified in many ways, the most useful depending on the extrinsic comparisons envisioned. Are we looking for an underlying structure? Then we see **(10)** in **(9)**. A surface structure? In that case **(11)**. The drainage pattern in **(12)** is further schematized in **(13)**, etc.

The simplification of a topography always depends on information extrinsic to it.

*The base map must be presented in low profile.* For the new thematic information to be as visible as possible, an efficient base map *excludes any sign which stands out.* Its lines should be unobtrusive; a line of dots (**3** and **4**) is the best formula. It is also the easiest and safest to draw. It furnishes a light outline which, unlike a fine line, does not risk disappearing when copied.

## *B.5.3. CARTOGRAPHY WITH ONE ORDERED CHARACTERISTIC*

The reader who has had the opportunity to see or use matrix constructions knows full well that to represent an order or a quantity in $z$, it is necessary to go from white to black. It is necessary to use either a *value* variation from light to dark, or a *size* variation from small to large. He also knows that the order of the visual variation must correspond with the order of the component. To conclude, he knows that all other visual variables transform a "seeing" map into a "reading" map.

**146**                                                          *Graphic constructions*

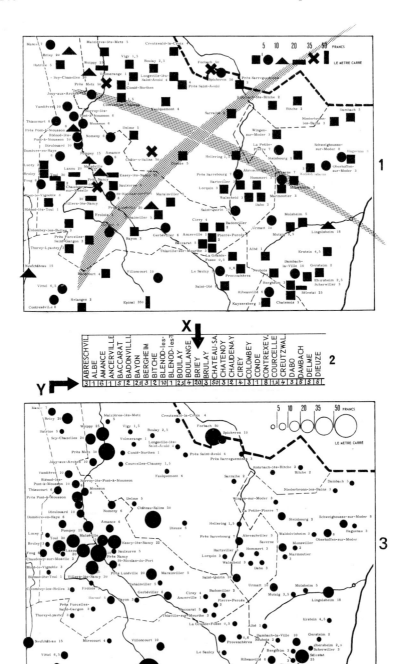

# Ordered networks: topography and cartography

### B.5.3.1. "Seeing maps; reading maps"

Accordingly, this reader quite naturally avoids the mistake of map **(1)** (reproduced by the million), which denotes the land prices by differences in shape, that is, denotes *quantities* by a variation *not visually ordered*. This map is a representation of data table **(2)**, containing two types of information:
Questions pertaining to $x$: what is the price of land in a given locality?
Questions pertaining to $y$: where is a given price? where is the expensive land?

**(1)** *is a* READING map. It provides answers to questions pertaining to $x$: what is the price of land at Briey? at Vittel? . . . provided that the legend has been learned. It does not answer questions pertaining to $y$: where is the expensive land? To find that out involves reading 117 signs one after the other. This map only gives the prices at an elementary level. At the level of the overall set, it shows only the points examined, with the quantities all thrown together. A "reading" map is not only a waste of time; it is a waste of information, and in fact *is usually not even read*.

**(3)** *is a* SEEING *map*. It allows us to see, that is, to immediately perceive the price distribution. It answers questions pertaining to $y$ and shows the groupings defined by both geographical and numerical proximity. It "regionalizes" the image and thus provides information about the entire set. *It also answers the elementary questions* pertaining to $x$. In a map with one characteristic, this overall level includes all the other information levels.

### B.5.3.2. How to avoid constructing reading maps?

Use SIZE or VALUE exclusively, since they are the visually ordered variables (p. 186). Texture, though also ordered, can only distinguish among a very small number of steps and remain efficacious.
The error committed in **(1)** is still widespread, especially due to the false aestheticism sought in color or to the unconsidered use of prefabricated screens. To avoid these mistakes it is sufficient to recall:
- that a legend is not needed for the perception of an order; any reader must be able to immediately class the signs from greater to less; the legend only serves to verbalize the increments between the steps.
- that this immediate classing obviously must correspond to the order of the component;
- that color variation is not visually ordered (see page 221).

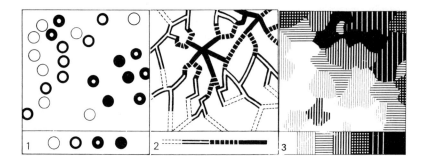

## B.5.3.3. The construction of "seeing" maps

*An ORDERED characteristic is transcribed by VALUE or SIZE.*

Basic rule: construct a scale with *visually equidistant steps* (p. 199)*. Indeed, steps which are not equidistant will obviously construct false visual groupings. Figures **(1)** to **(5)** show the simplest transcriptions for each "implantation." For implantations by points **(1)** or lines **(2)** do not exceed four or five steps of value. For a larger number of steps, use size **(6)** and **(7)**. For implantation by area, arrangement **(3)** can be created manually with commercially available screens**. Traditional computer print-outs generally produce a very rough image **(4)**. Photocomposers and special screens (p. 207) produce much more accurate and reliable images, such as **(5)**, which can accommodate more than fifteen steps and are progressively replacing images such as **(4)**.

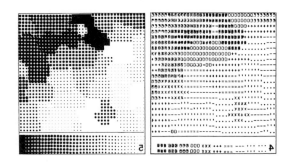

*Ordered networks: topography and cartography* **149**

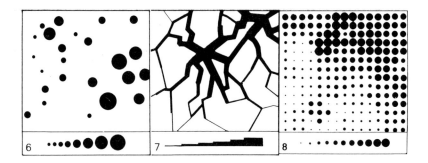

*A QUANTITATIVE characteristic is transcribed by SIZE.*

Basic rule: *proportionality between visual distances and quantitative distances.* In fact, the aim of the representation is to make the groups created by the quantitative distances appear. The simplest graphic arrangements are:
- for implantations by points **(6)** or area **(8)**: circles of proportional area (p. 205);
- in linear implantation **(7)**: proportional widths.

Let us remember that a quantitative characteristic can be transformed into classes of quantity and be represented as in **(1)** to **(5)**. However, in these cases the proportions disappear, taking along with them the distances between the numbers, and the resultant groupings. Only the classes constructed *a priori* remain.

---

*And *S. G.*, p. 75.
**It is indeed regrettable that these screens are not conceived in a logical way, that is, for the two essential modes — points and lines — 1st a value variation (with visually equidistant steps, not ordered 10, 20, 30 . . . %); and 2nd a texture variation for each successive step. 70% of the plates are either irregular or useless. They merely contribute to graphic error, for example, the fifth step in **(3)**.

150 *Graphic constructions*

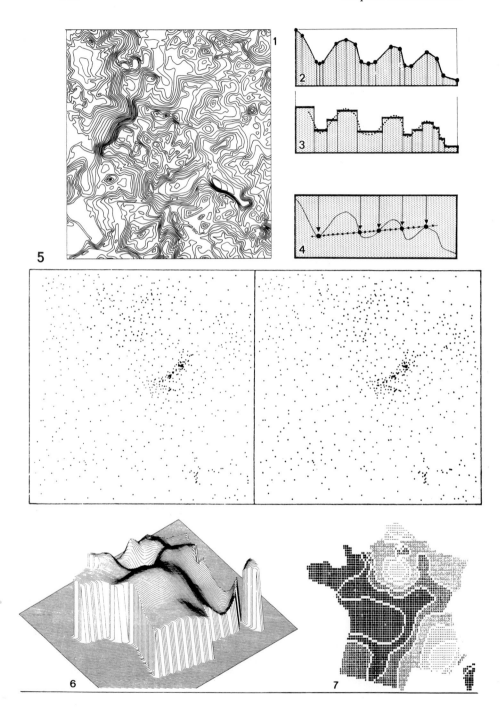

## B.5.3.4. READING MAPS

*Isarithms (or iso-lines)*
Isarithms precisely define a quantity at each point on the paper. But, except for drawings supplemented by a value variation (7), the curves construct reading maps (1), which is what we generally ask of them, but not "maps for seeing quantities." We see only the slopes (1). In order to properly construct these curves we must distinguish two types of information:
a) When the statistical surface is entirely known, as, for example, with land surface. In this case we choose a necessary and sufficient number of control points to reconstitute all the points (2). When information by area* is involved, as with population density per community, we must first smooth the surface (3), which brings us back to the previous problem. Ambiguity here is simply due to human error.
b) The surface is invisible, as with a geological surface that we are trying to reconstitute by sampling. Here there will always be ambiguity(4); hence the use of calculations that bring to bear on each unknown point the structure resulting from the set of known points through smoothing, a structure specific to the phenomenon, or a coefficient of its variances. Different models can be used. The French model** supplies two maps: surface curves and isovariance curves, the latter indicating the probability which can be attributed to this surface.

*Stereoscopic pairs (5) and anaglyphs*
They produce the best visible reconstitution of a surface defined by curves or control points. They are easily constructed by computers, but the comparison of several surfaces is very complex, if not impossible. A stereoscope must be used to discover the surface defined by the pair in (5).***

*Use of perspective*
Anything that interrupts a plane is not a map; $xy$ is not homothetic to the geographical constant, and the image is not universal. How much time would it take to specify the difference in information**** between (6) and (7)?
A "perspective" could be a map, but it is curious to note the extent to which cartographers and computer programmers believe that relief is due to planar deformation. In fact it results from a deformation of details. Planar deformation merely creates the illusion of perspective, which is not the same as illustrating relief. Incidentally, to create this illusion one need only cut the map into a quadrilateral whose boundaries converge towards the top of the figure (S.G., p. 380).

In conclusion, curves without value variation, stereoscopics, and perspectives do not produce an image of the overall set. Imagine for a moment having to class by similarity 200 maps prepared according to one or another of these formulae. It would be practically impossible, whereas this operation could be carried out in a few minutes with maps where $z$ is transcribed by variable shading running from white to black.

---

*In this case, the English cartographical term in isopleths.
**G. MATHERON,"Le Krigeage Universel,"*Cahiers du Centre de Morphologie Mathématique de Fontainebleau.* France (1969).
***From M. GUY. *Geoforum,* 3 (1970).
****From J.-C. MULLER. *La cartographie thématique aux U.S.A.* Paris: Roneo.

## B.5.4. CARTOGRAPHY WITH SEVERAL CHARACTERISTICS

In these cases cartography must pay for its specific properties. One might say that in a data table, $x$ represents the geography, $y$ the characteristics, and $z$ the quantities. But in cartography, $x$ and $y$ represent the geography and $z$ the quantities. Therefore the map lacks one dimension for representing additional characteristics in a single image.*

### B.5.4.1. Elementary information or overall information about the set?

Take the distribution of the labor force into three sectors: I — Agriculture, II — Industry, III — Tertiary, T — Total, for the ninety French departments. We know that a data table is constructed to supply overall information about the set: how do the $x$'s and $y$'s fall into groups? In other words: what is the geographical form of each characteristic and which characteristics have the same form?

*A superimposition* **(1)** *answers the elementary question:* at a given place, what is there? This question corresponds to the $x$ entry in the table. But there is no visual answer to the question: where is a given characteristic? i.e. to the $y$ entry in the table. To find this answer we would have to individually analyze the ninety departments. Map **(1)** is a figure made up of *ninety images* which would have to be memorized. In short, a useless map. The four maps in **(2)** are impossible to see in **(1)**.

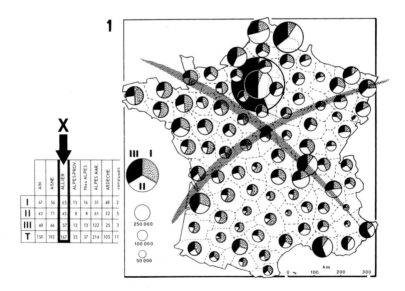

*Ordered networks: topography and cartography* **153**

*The collection* **(2)** *answers questions about the overall set:* how do the characteristics fall into groups? Indeed, each map immediately answers the question pertaining to *y*: where is a given characteristic? This results in the perception of the similarity between II, III and T as compared to I.

*We may conclude that the main problem in cartography with several factors will vary, depending on whether or not a reading of the overall set is necessary, that is, whether we are to answer the question pertaining to y*: where is a given characteristic located?

Such a question is *pertinent* when we must compare numerous characteristics, determine geographical correlations, or define regions and their boundaries. This information is at the level of the whole set. Reading the overall set is necessary in such cases, and the solution is a collection of maps.

Such a question is *irrelevant* when we are simply looking for a precise point on the map, the location of a village, a pipe line, a community boundary or the best route to take. In such cases the sought-after information is at the level of the single element. *Elementary reading suffices:* the solution is a superimposition.

But if the map were simplified, could we not, perhaps, respond at both levels of information!

*Differential characteristics, different professions for example, construct a data table with several yes/no rows. It therefore constructs a map with several yes/no characteristics.

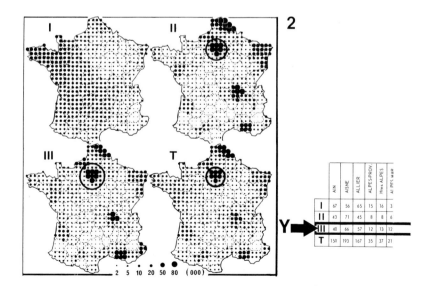

## B.5.4.2. Simplified information or comprehensive information?

Let us consider map (3). It superimposes the geographical arrangement of II, III and T onto I. This map is easy to read and provides an immediate answer to all questions, but at the cost of considerable simplification of the data. Map (3) does not allow us to reconstitute the initial information, that is, the comprehensive information transcribed in (1) or in (2), and it offers no means for discussing the level of simplification.

*We may conclude that the main problem in cartography with several factors will vary depending on whether or not it is necessary to transcribe the comprehensive information, that is to say, depending on whether we are at the stage of looking for a useful simplification or at the stage of communicating this simplification.*

In short, cartography with several factors depends a) on the level of information being sought and b) on the point in processing: research or communication. These two alternatives intersect and define, for specific data, the four forms of cartography:

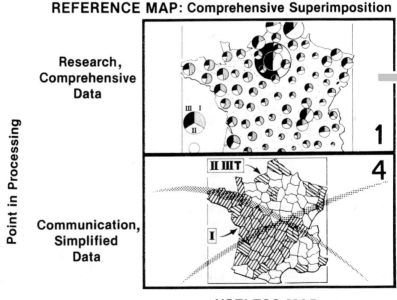

**(1)** comprehensive superimpositions, readable only at the elementary level and useful for reading what there is at a precise place. This is the purpose of architectural plans, topographical maps, road maps, "complex" maps, nomenclature maps, etc... *These are reference maps or mark maps,* useless for relationships involving the entire set.*

**(2)** a collection of comprehensive maps, one per characteristic, readable at the overall level and useful for defining regions and geographical correlations. *These are processing maps.*

**(3)** depicts the simplified superimposition, readable at all levels and useful for memorization and teaching; with these we must keep in mind that the level of simplification is always open to discussion. *They are communciation maps.*

**(4)** represents those maps which offer neither comprehensiveness nor overall reading. Accordingly, these are useless maps.

---

*In the example chosen, it is obvious that for elementary reading **(1)**, the table of numbers is more precise and efficacious than the "chart-map" **(1)**. As a rule, *"chart-maps",* that is *diagrams scattered over a map,* are totally useless for reading overall relationships and insufficient for elementary reading. That is why map **(1)** on p. 152 has been crossed out.

## INFORMATION
### Overall Reading: "Seeing Map"

### PROCESSING MAPS: Collection of Comprehensive Maps

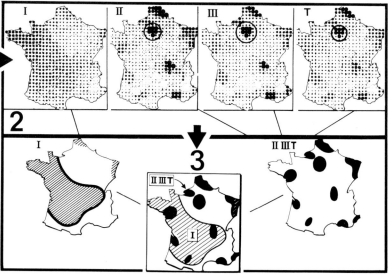

**COMMUNICATION MAP: Simplified superimposition**

# Graphic constructions

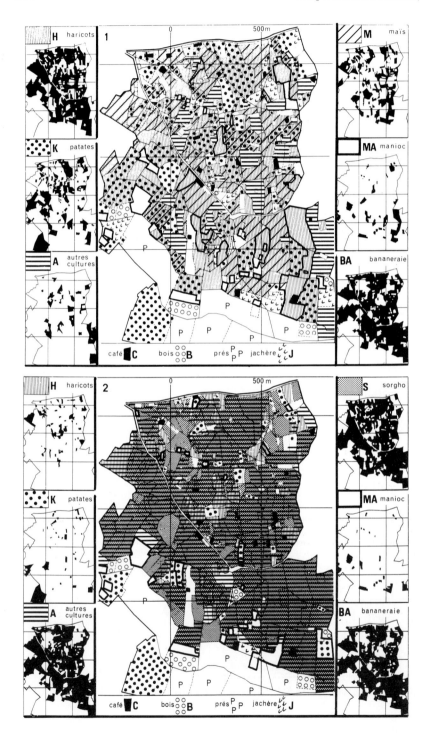

*Ordered networks: topography and cartography* 157

## B.5.4.3. Using cartography to publish comprehensive information

Thus it is not possible, with complex information, to represent several characteristics in a comprehensive way on a single map while simultaneously providing a visual answer to our two types of question. When we ask a cartographer to produce a map for several factors, he must first of all determine the type of question which will be asked.

*When both types of question are pertinent*, as is generally the case, there is only one satisfactory solution: making several maps.
1st. Making a superimposition map for answering questions in $x$: what is at a given place?
2nd. Making one map per characteristic for answering questions in $y$: where is a given characteristic?
Two examples follow.

*Crop Cycles in an African Region**

*Information:* cultivated field with seven crops or combinations of crops.** Two seasons: (1) first crop cycle; (2) second crop cycle.
*Transcription:* one superimposition map per cycle, superimposing the crops. One map per crop in the margins of the superimposition maps.
*Result:* if we remain bound to only the two superimposition maps, a long analysis would be necessary in order to define the changes which occur between the two cycles and the differences within the region. The collection of maps by characteristic makes these changes obvious and enables us, moreover, to show a noticeable difference between the region's northern and southern parts.
Note that the maps by characteristic:
- do not pose a representational problem and can be quite small;
- do not require additional work, since the cartographer must produce them anyway, in order to study the signs to be used in the superimposition or in order to draft the plates in corresponding colors;
- do employ a reference grid, allowing precise comparisons among all the maps.

*From L. MESCHY. *Kanserege, une colline au Rwanda.* Thèse de Doctorat. Paris: EPHE, 1974.
**H. Beans, K. Sweet Potatoes, A. Other Crops, C. Coffee, B. Woods, P. Meadows, J. Fallow land, M. Corn, MA. Manioc, S. Sorghum, BA. Banana-Plants.

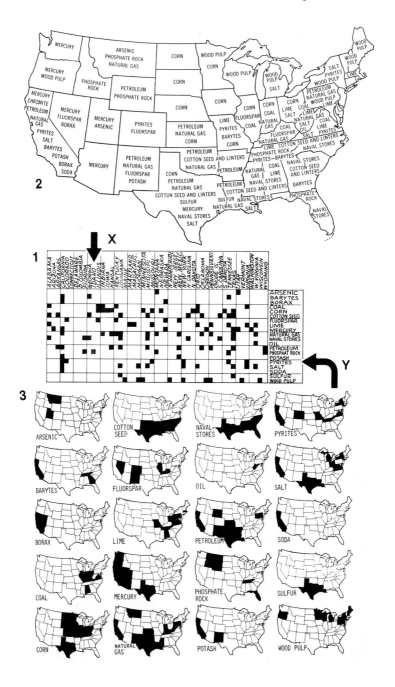

## Ordered networks: topography and cartography

*Raw Materials for the US Chemical Industry**

*Information:* along *x* the States of the USA; along *y* twenty raw materials **(1)**.
*Transcription:* a map and written notations **(2)**. A perfect example of a reading map.

*Result:* this map immediately answers the question: what is there in a given state? But what happens if we have the misfortune to ask: where is the petroleum? We have to read every word on the map in order to obtain a correct answer, that is to avoid omitting any producing state. At the same time we would have to memorize production by state, which very quickly becomes impossible. And this operation would have to be repeated for each product! Map **(2)** answers elementary questions in *x* visually, but provides no visual reply for questions in *y*. These answers may only be supplied by a collection of maps **(3)**.

*A collection is reclassable.* It may be reclassed in different ways to correspond to a particular research topic. In **(3)** it is classed alphabetically. A geographical classing **(4)** enables us to discover regional traits. A comparison of the maps involving a given state would point out rival industries. A classing by types of product would be useful in teaching, etc.

In conclusion, in the publication of maps, the graphic transcription must answer all pertinent questions. The rule is as follows:
*All comprehensive cartography involving a multi-row data table must use not one but two constructions: a superimposition to answer the questions in* x *and a collection to answer the questions in* y.

*From *Chemical Facts and Figures*, 1940.

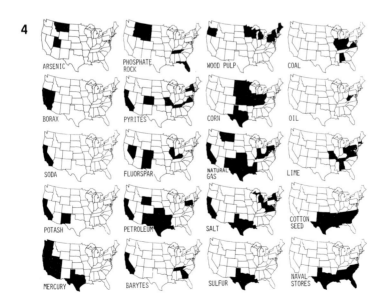

# Graphic constructions

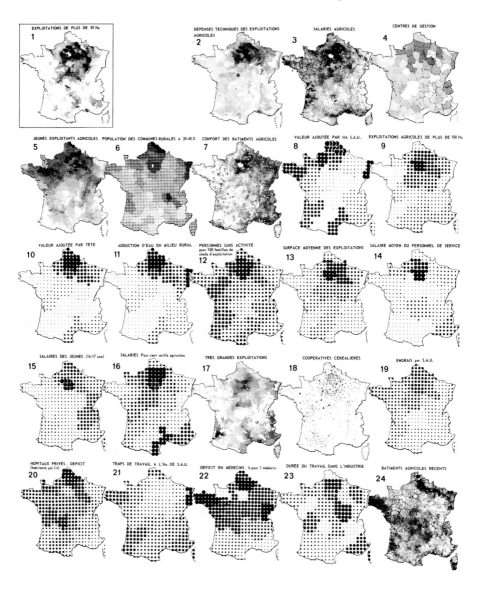

1. Farms larger than 50 ha. - 2. Technical expenses for the farms. - 3. Salaried farm workers. - 4. Administrative centers. - 5. Young farmers. - 6. Population of townships 20-40% rural. - 7. Comfort of agricultural buildings. - 8. Added value per ha. - 9. Farms of more than 100 ha. - 10. Added value per head. - 11. Water supply in rural areas. - 12. Percentage of non-working persons in farming families. - 15. Average salary of young peo-

*Ordered networks: topography and cartography*                                        **161**

## B.5.4.4. Cartography and data processing

Cartography is above all a means of data processing, independent of problems involved with publication. It can serve either in the discovery of characteristics corresponding geographically to a given characteristic or in the discovery of a geographical distribution defined by a given set of characteristics (the "synthesis" map.)

*The discovery of characteristics corresponding geographically to a given characteristic.*

When we study a question — for example, a question connected with large farms (first map, top left) — the essential problem is to discover which characteristics have a relationship of causality or coincidence with the phenomenon being studied. The number one problem involves imagination; measurement of coincidence can only come later.

If we have a bank of mapped information, it is easy to discover — in a short time (in this case three quarters of an hour) from a sizable collection (in this case 600 maps) — a series of maps whose distribution is directly or inversely related to the distribution in question. The phenomena represented by these maps are likely to have a relationship of causality or coincidence with the phenomena under study.

For the human imagination, always too limited, always curbed by socio-cultural contexts, map collections present possibilities as vast as the data bank is large. Visual selection is faster and better than any automatic selection, since it permits from the outset a variety of nuances beyond the capability of any computer. But its cost in terms of time only pays off with "seeing maps." "Reading maps" make the operation impossible.

*Properties of a map collection*

- The collection of comprehensive maps *does not involve problems of generalization*. Indeed, the eye immediately sees a shape, whatever its complexity. Each map can thus carry an impressive amount of data, as with the twenty-five million buildings in Poland on a scale of 1/2 M (F. UHORCZAK*). But at the same time, the eye is free to focus on any level

<small>ple (14-17 years). - 16. Percentage of salaried workers to agricultural workers. - 17. Very large farms. - 18. Grain cooperatives. - 19. Fertilizer per agricultural area. - 20. Shortage of public hospitals. - 21. Working time per ha. - 22. Shortage of doctors. - 23. Length of time in industry. - 24. Recent agricultural buildings.
*S.G., p. 177 and 318.</small>

*Graphic constructions*

1. DOCTORS  2. MIGRATIONS  3. FARMERS

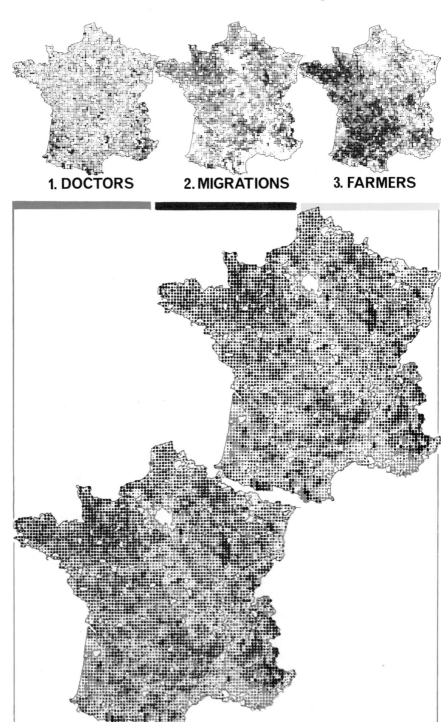

## Ordered networks: topography and cartography

of an ordered or quantitative variable and is thus free to "generalize," that is to regionalize, as it pleases.

- *Such a collection accommodates substantial reductions* for the reason mentioned above; accordingly, it enables us to assemble easily classable collections.
- *It excludes problems of standardization.* In effect a variation from white to black is sufficient, and for quantitative variables, our studies all point to one method: proportional circles, weighted either by points or by regular pattern.
- *It can be easily automated,* so, it is not unwieldy, for all its flexibility and speed. Nor does it pose particular problems in drafting.
- *It poses no particular problems in publishing,* and we can even display a map on a cathode screen and make it disappear without setting it on paper, since we can always make it reappear.
- Finally, and above all, *it is not limited in number of variables.* Remember that modern science poses the problem of correlating an ever increasing number of variables, hundreds and soon thousands. This alone would be sufficient to explain the development of collections of maps each having one variable.

*Realization of a "synthesis map." A—Trichromatic procedure*

The collection of maps does not answer the question "what is there at a given place?" But maps with the same scale can be superimposed three by three. It is sufficient to transcribe them on three different color films: cyan-blue, yellow, magenta-red\*. With "Colorkay" films, for example, one can carry out this transcription oneself.

Take maps **(1)**, **(2)**, **(3)** on the facing page, depicting percentage of population; visual comparison of the maps is difficult. But the superimposition **(1 + 2)** shows immediately that there is a negative correlation between doctors and migration. A positive correlation would produce a violet map, whereas here we only see red or blue.
The trichromatic superimposition **(1 + 2 + 3)** enables us to demarcate the regions defined by the combination of the three characteristics. We can thus verify that the eye is capable of making a nearly instantaneous synthesis of about $7000 \times 3 = 21{,}000$ items of information.

---

\*Any other color, for example any other red or blue, makes the synthesis impossible (p. 217).

*Realization of a "synthesis map." B — Cartographic procedures*

Consider the ecological planning study for the West Toulon area.* It involves directing the growth brought about by the construction of a motorway and the arrival of water from the Provence Canal as well as protecting the landscape of Provence, with aesthetic values having overriding importance. In essence, it is a matter of defining the best agricultural areas, the areas to protect, the areas suitable for urbanization, and those where these different purposes conflict.

*PHASE I: collection of mark (reference) maps.* We construct a map for each useful characteristic: soil value, water resources, forest cover, slope, sunshine, frost, microclimates, scenic beauty (map **1**), historic sites, etc.
*PHASE II: intermediate synthesis maps.* Constructed from the mark maps, map (**2**) defines the areas to preserve for the quality of their woodlands and their situation in the countryside. Map (**3**) defines the best areas for the development of agriculture.
*PHASE III: general synthesis map.* Map (**4**) results from the superimposition of the intermediate synthesis maps plus the urbanization map. It indicates the best utilization of each area: A. areas to protect, B. most suitable for agriculture, C. sites for urban development, D. areas where purposes conflict.

Several observations are to be retained from this cartographic procedure:
- Maps with several characteristics are excluded from phases I and II. Comparisons are made with maps having a single characteristic represented in several degrees of value.
- One characteristic may register on several synthesis maps. For example, slope, sunshine, frost.
- The general synthesis is the only map with several characteristics ABCD. But each characteristic may have several degrees of value, which raises difficult design problems.
- The general synthesis map results from various compromises, which may be called into question, necessitating the construction of a new map. If manual cartographic procedure becomes too cumbersome, automation affords the solution, permitting us to quickly reconstruct the geographical distribution defined by a given compromise.

*From M. FALQUE, A. GALAND and J. TARLET, in *Moniteur des Travaux Publics*. Paris (March, 1975).

# Ordered networks: topography and cartography

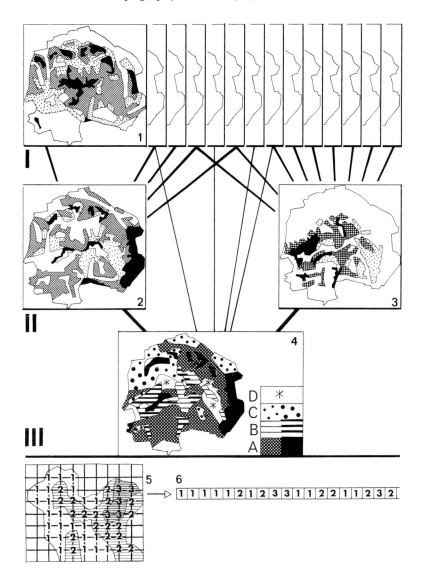

– To put manual maps into a computer, one must "digitize" them. A grid is placed on each map (**5**). Each point is matched with its value. Each map then becomes a row (**6**). The set of maps thus constructs a matrix, to which any multidimensional processing method can be applied.

**166**                                              *Graphic constructions*

*Realization of a "synthesis map." C — Matrix procedures*

Consider having to choose ten representative townships from 100 in the Ardennes department of France, in order to proceed to detailed surveys.* To be certain of the best possible representation, ninety-two characteristics, chosen by specialists from the department, are taken into consideration. The information constructs a matrix 92 × 100, which can be reduced mathematically or graphically.

Graphic permutations construct a first "system" **(1)**, from which we obtain interpretation matrix **(2)**. It makes a classing appear that goes from private commercial services (at the top) to agricultural operations (at the bottom), passing through degrees of commercial activities and cooperative farming. This classing thus reveals *"degrees of urbanization"* and accounts for more than half of the characteristics. These degrees are denoted by legend **(3)**, which leads to.map **(4)**. The characteristics not appearing in this system are processed later.
They construct a second "system" **(5)**, and the interpretation matrix **(6)** reveals a new classing whose main characteristic is the size of farms. Denoted by legend **(7)**, it produces map **(8)**.
The superimposition of the two maps **(9)** provides the general synthesis from which we can choose ten townships representative of the types thus discovered.

In relation to the preceding procedure, this example is characterized by a substantial reduction in the *a priori* part of the operation. The intermediate syntheses **(4)** and **(8)** stem from the set of characteristics and are only identified after analysis. In the previous example, however, the maps of agriculture, protections and urbanization were constructed *a priori*. But are these traditional definitions really meaningful in the modern world? In certain cases, is it not preferable to discover these new "factors" resulting from a maximum number of original characteristics, and thereby represent lived reality more accurately, even though these factors may not as yet have risen into the public consciousness. Other examples of this procedure will be found on pp. 48, 81, and 85.

---

*From J.D. GRONOFF, "La zone herbagère des Ardennes," *Etudes Rurales*. Paris: Mouton, 1971.

## Ordered networks: topography and cartography

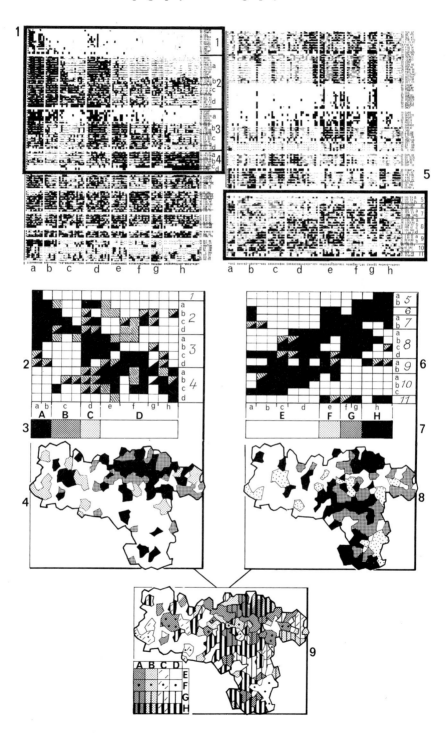

## B.5.4.5. Visual selection in superimposition maps

Superimpositions of characteristics pose a problem of visual selection. It is common practice to resort to color differences, but this seemingly easy solution is not without serious inconveniences (p. 222). Is it better to publish a black and white map or *not to publish* one in color? Moreover, we discover that in many cases other visual variables yield as good a selection while avoiding all the pitfalls involved with color. It is sufficient to keep in mind four recommendations, derived from the properties of visual perception.

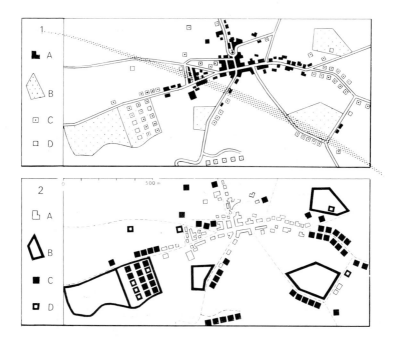

*a — Highlight the thematic information, not the background.*
In this case we wish to show the extent of new construction in a village. By habit, the map's designer **(1)** gives this village the appearance of a quiet rural agglomeration, whereas in fact, the village is in the process of becoming a residential suburb **(2)** with all its characteristics!
A — Construction before 1962   B — Sub-divisions
C — Construction after 1962    D — Applications for building permits

### Ordered networks: topography and cartography

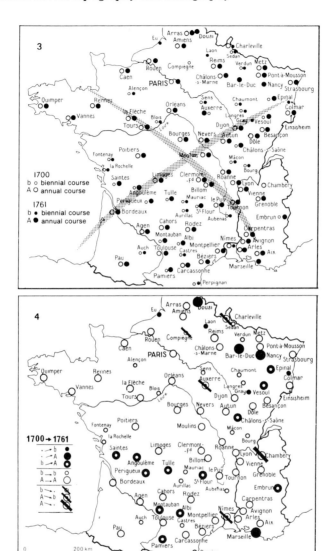

*b* — *Replace two maps by the map of their difference.*
It is certainly difficult to see the development of Chairs in Physics between 1700 and 1761 on map (3), superimposing both situations. Contrastingly, a precise study of the cases leads to the legend and map (4), which immediately separates out areas of stability, progress and decline. Note the use of an oriented sign.

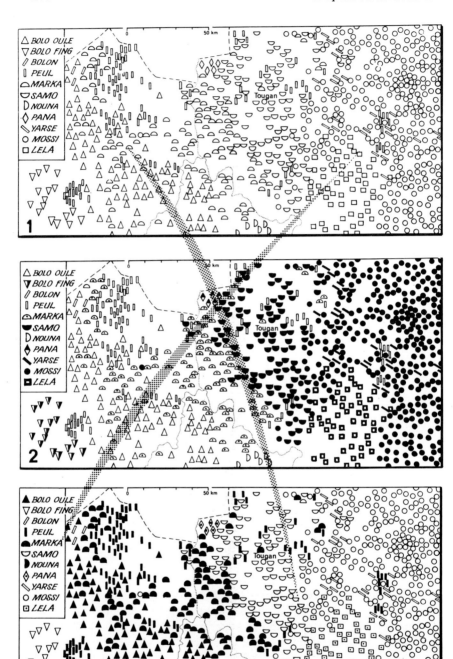

# Ordered networks: topography and cartography

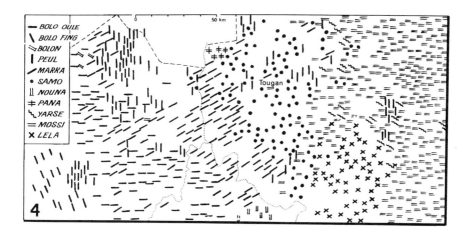

*c* — *In implantation by point, replace a variation in shape by a variation in orientation.*

Take eleven ethnic groups from the Tougan region (Republic of Upper Volta). Map **(1)** differentiates the ethnic groups by shape. Population density is visible. Maps **(2)** and **(3)** utilize shape and value. The population density varies with the whim of the designer.

But in implantation by point, a variation in orientation **(4)** enables us to preserve our perception of density, since all the signs are of equal value, while offering a much greater selectivity than shape variation. One need only look for the distribution of a given ethnic group in **(1)** and in **(4)** in order to realize this. Furthermore, map **(4)** may be drawn rapidly, by anyone, whereas **(1)** would be time-consuming and difficult to carry out.

A much more complex example is given on the following page. It shows the different trees and bushes in an African region. Observe the ease of isolating an oriented sign on page 173, in comparison to its analogues on page 172.

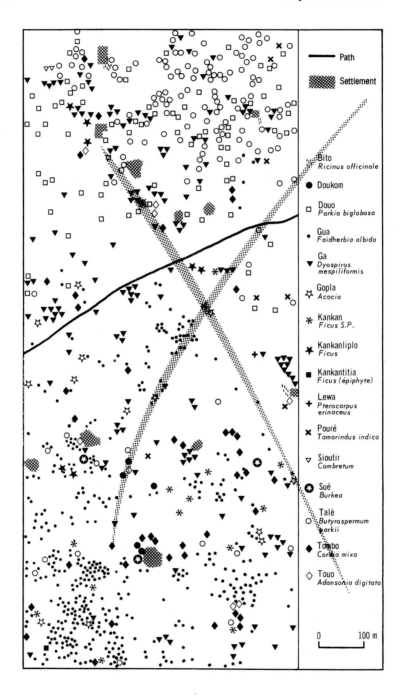

# Ordered networks: topography and cartography

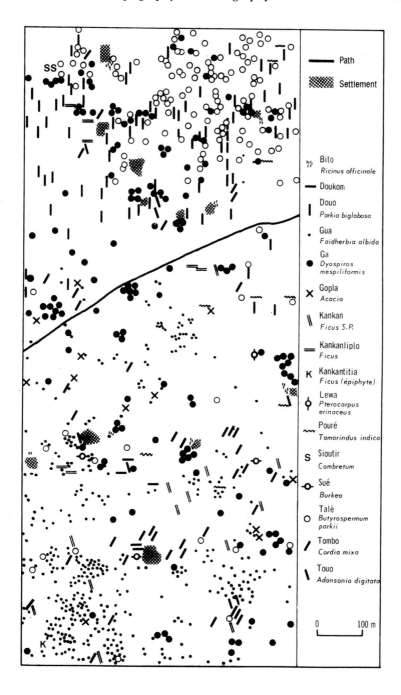

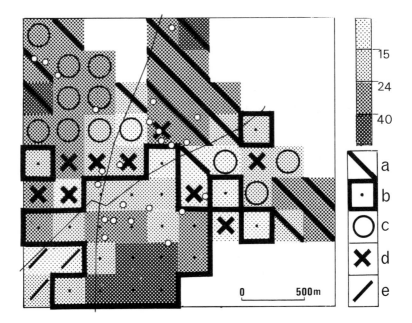

*d* — *Use a difference in texture\* and contrast point/line/area.*
*1st example.* Take a superimposition of tree density (number of trees in a square whose sides are about 250m) upon types of trees (a, b, c, d, e). Since the study is concerned as to the dominant type, the types themselves are not superimposed on each other.

*Distinguishing between density and types* is accomplished by a difference in texture. The densities, in very fine texture, show through the types, in very rough texture.

*Distinguishing among types* is accomplished by a combination of shape, orientation and size, and the homogeneous distribution of *b* (lower left) allows us to use a single "wrap-around." The white points represent settlements.

*2nd example.* Take a superimposition of several regional characteristics— rainfall, dry seasons, forests—in Gabon, upon two linear characteristics: axes of maximum or minimum rainfall.\*\* All are superimposed. *Distinguishing between rainfall (A) and the other characteristics* is accomplished by the contrast between areas and lines.

\*\*From G. SAUTTER, *De L'Atlantique au fleuve Congo. Une géographie du sous-peuplement.* Paris: Mouton, 1966.

*Ordered networks: topography and cartography*

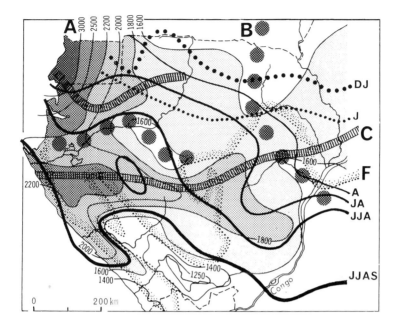

Distinguishing minimum (B) or maximum (C) rainfall axes from dry season lines is accomplished by the size/value combination. (B) and (C) are gray, but of larger size.

*Distinguishing between minimum (B) and maximum (C) axes* is accomplished by texture, nonexistent in C (continuous line), and coarse in B (line of points).

*Distinguishing between boreal dry season (DJ and J) and austral dry season (A, JA, JJA, JJAS)* is accomplished by texture. DJ and J signify December (D) and January (J), the months with less than 100 mm of rainfall. In the austral dry season, June (J), July (J), August (A), and September (S) are the months with less than 30 mm of rainfall.
The boundary of the forest (F) is denoted by a clear gradated line of medium texture.

*Final recommendation.* Whenever several characteristics are to be superimposed, the least bad representation depends above all on the distribution of the characteristics. There is only one solution: undertake several trials and remember that we should not ask the cartographer to be a designer. We ask him to resolve problems of relationships among sets.

# C. THE GRAPHIC SIGN SYSTEM
## (A Semiological Approach to Graphics)

Approaching graphics as a sign system enables us to set the preceding observations into a broader context.
*Definition.* Graphics involves utilizing the properties of the plane to make relationships of resemblance, order or proportion among given sets appear. Graphics is the monosemic level of the world of images.
"Graphics" designates the sign system. A "graphic" designates any construction produced within this system, whether a diagram, a network or a map.

## C.1. SPECIFICITY OF GRAPHICS

### C.1.1. Graphics and pictography: two domains of signification

*To perceive a pictograph,* a road sign **(1)** for example, requires only a single stage of perception: what does the sign signify? Stop! All the useful information is perceived. The aim of pictography is to *define a set* or a concept.

*To perceive a graphic* requires two stages of perception: 1st: What are the elements in question? 2nd: What are the relationships among those elements.

Drawing **(2)**, for example, is useless; the second stage of perception is missing. But drawing **(3)** is also useless; the first stage of perception is missing. The aim of graphics is to *make relationships* among previously defined sets appear. A graphic is always the result of a double-entry table. A pictograph, never! Otherwise it would be a graphic.

1

2

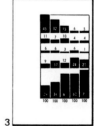
3

## Specificity of graphics

*The two stages of graphic perception*
*In the first stage, external identification,* the observer isolates given sets, for example a set of countries or products (2), from among all possible sets. This is the "sign" stage, that is the stage of verbal conventions or conventions more or less analogous to pictography. Before the infinity of possible significations, a sign is always accompanied by a certain ambiguity. It is always subject to discussion. It is polysemic. However,

*without external identification a graphic is useless. The external identification must be immediately legible and comprehensible.*

The perception of any graphic begins with the question: what do the *x*, *y* and *z* dimensions of the image represent? Writing these definitions in an illegible way, in letters that are too small (2), or placing them on other pages shows that the author is confusing graphics with pictography and that he has never used graphics for interpretation and decision-making.

*In the second stage, internal identification,* the observer discovers relationships. This is the true domain of graphics. The transcription of relationships does not utilize "signs"; it utilizes only the *relationship between signs*. It utilizes visual *variations*. Graphics denotes a resemblance *between* two things by a visual resemblance *between* two signs, the order of three things by the order of three signs. But be careful! to denote an order by a resemblance is not a convention; it is a fallacy. The transcription of relationships is without ambiguity. Visual variables are monosemic. Graphics has absolute natural laws. It is not "conventional."

*Resemblance, order and proportion are the three signifieds in graphics. These signifieds are transcribed by visual variables having the same signifying properties.* *

How do we represent a factory? There is an infinite number of "good" representations. The choice is an art. That is pictography.

---

*On the basis of these remarks we may say that graphics (and therefore cartography) does not conform to a polysemic schema: sender ⟷ code ⟷ recipient, but to a monosemic one: operator ⟷ three relationships. Sender and recipient are identical in terms of their goals: understanding relationships. There are only "operators." And in fact, we note that unlike the monosemic schema, the polysemic schema does not enable us to solve graphic problems.

Factory A employs twice as many workers as factory B. There is only one single representation: show that A is twice as large as B. This is not an art since *there is no choice*. This is graphics. *The graphician* is bound by the properties of the visual variables because these variables are not all visually ordered or proportional.

Graphics and pictography are two languages, fundamentally different in objective. Confusing them, which leads to fallacious conventions, is the source of the most blatant graphic errors.

### C.1.2. Mathematics and Graphics: two systems of perception

*Analogy between graphics and mathematics*

Both graphics and mathematics start from sets previously defined in verbal or visual terms and not subject to further discussion. An equation only makes sense when the unique signification of each sign is recognized.

A graphic only makes sense when the given sets proposed for study are recognized. Once these sets are recognized, graphics and mathematics are only concerned with the three basic relationships to which any observation can be reduced.

| | | |
|---|---|---|
| Defining a set | LANGUAGE | PICTOGRAPHY |
| Making the relationships among defined sets appear | MATHEMATICS | GRAPHICS |

*Difference between graphics and mathematics*

Graphics utilizes the three dimensions of the image. It is a spatial "sign system" independent of time.

Mathematics utilizes the two dimensions of sound. It is a linear "sign system" defined by time.

Remember that written notations of music, words and mathematics are merely formulae for the memorization of fundamentally auditory systems, and that these formulae do not escape from the systems' linear and temporal character. The ear can hear an equation on the telephone; it cannot hear a map.

| LANGUAGE MATHEMATICS | PICTOGRAPHY GRAPHICS |
|---|---|
| AUDITORY PERCEPTION | VISUAL PERCEPTION |
| 1 variation in sound (or shape) across 1 variation in time (t) | 1 variation in mark (Z) across 2 dimensions of the plane (XY) |
| 1 dimension + t | 3 dimensions XYZ |

*Specificity of graphics*

In a moment of perception, the ear perceives one sound. In a moment of perception, the eye perceives the relationships among three sets. No other system of perception possesses this property, and it seems indeed that logic* is based on the three dimensions of visual perception.

To utilize the linearity of time in order to represent a set of relationships is the difficult problem that auditory systems must resolve.

*To utilize the immediacy of the image in order to represent a set of relationships is the main problem of graphics.*

We utilize graphics to save time and consequently memory; in order to SEE, that is to perceive immediately. Accordingly, a graphic which must be READ, that is perceived over time, does not solve the problem. Moreover, we observe that such a graphic is usually not even read. The reader prefers the written text, since it generally yields a much better ratio of information received to time spent.

*Place of graphics within the basic sign systems*

Two facts separate the basic sign systems:
- a defined set and a set to be defined distinguish between the two domains of signification;
- the eye and ear distinguish between two separate systems of perception. These two facts intersect and enable us to construct the table below.

| | | SYSTEM OF PERCEPTION | |
|---|---|---|---|
| | | 👂 | 👁 |
| SIGNIFICATION ATTRIBUTED TO PERCEPTIONS | pansemic | MUSIC | NON FIGURATIVE PICTOGRAPHY |
| | polysémic | LANGUAGE | PICTOGRAPHY |
| | monosemic | MATHEMATICS | GRAPHICS |

*Mathematics and graphics exclude polysemic signs, by only considering relationships among previously defined elements, no longer subject to discussion. This monosemic property then enables us to discuss the set of elements and to connect the propositions in a sequence of truths which can become undebatable, that is, logical. Monosemy is a fundamental condition of logic, though also its drawback. Monosemy only exists within a defined set. But there is an infinite number of defined sets, howsoever large they may be individually. Therefore, logic amounts to a sequence of rational moments, always immersed in the infinity of the irrational.

## C.2. THE BASES OF GRAPHICS

### C.2.1 A graphic! To do what? Levels of information

Information is a relationship between two elements or between two sets of elements. In graphics, it is the answer to a question. What are the questions we can ask of a data table?

The hotel example (p. 11) reminds us that there are *two types of question*; questions pertaining to *y*: *older guests*, when are they most numerous? But this example also shows that for each type we must distinguish three levels of question, that is *three levels of information*.

*The elementary level:* in a given month or a given category, how much? The answer is provided by the number in the cell. This is elementary information, and information processing must struggle against it. In fact, our memory cannot retain the multiplicity of elementary data. We must reduce this multiplicity, discover similar elements, group them, and class them. These are the prerequisites to understanding and deciding.
*The overall level* of the entire set is thus the true objective. What are the different periods that the entire set of characteristics constructs? This is the level necessary for decision-making.
*The intermediate level* comprises all the subsets included between these two extremes: i.e. what are the characteristics of summer? The intermediate level enables us to analyze the overall relationships.

*The processing of a data table reduces the* x *and* y *sets to the groupings constructed by the* z *numbers. Understanding is the discovery of these groups and the relationships they engender. This is the main problem in information processing.*

### C.2.2. Properties of the image

Faced with this problem of data reduction, the image possesses three properties.

*The image has three dimensions: x, y,* and *z.* Any point in an image **(1)** can be perceived as the correspondence between a position along *x*, a

*The bases of graphics*    **181**

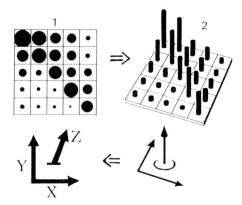

position along *y* and an elevation in *z* (**2**). The set of points can be perceived as the set of correspondences among three dimensions *x, y,* and *z*.

*A meaningful form is independent of the number of its parts.* The eye SEES a tree whatever the number of leaves. The image accommodates a very large number of elementary data, limited only by the "threshold of differentiation" and by material conditions. This is how the eye can immediately see the difference in distribution between two maps each comprised of twenty-five million data items.

*The eye can focus either on an element, or on a group or on the overall image.* It can SEE a leaf, a branch or the entire tree, and it sees the relationships they engender. The eye perceives the three levels of information spontaneously. With respect to the plane, the eye possesses the property of ubiquity.

The eye perceives subsets. It compares them and discovers differences and resemblances. It perceives the resemblance between two rows of a matrix, but it also perceives their differences. If this difference signifies nothing, it is possible to *eliminate this meaningless difference* by bringing the two rows together. This is the aim of permutations.

*The* x y z *image can represent the quantitative relationships between two sets of elements. It can also,* provided it is physically permutable, *reduce the multiplicity of elementary information to the groupings constructed by the data. It enables us to discover information about the entire set.*

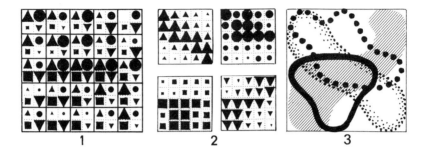

*But the image has only three dimensions.* Can we superimpose several images? Can we superimpose several different characteristics on a map **(1)** and still retain the properties of the image? No more than it is possible to superimpose several photos on the same film. The image has only three dimensions. This is its limit. The fourth dimension is time. It is time which is needed to discover in **(1)** the four images in **(2)**.

In a superimposition the eye only sees the form created by the sum of the characteristics. If this sum is not meaningful, the overall form is without signification. To understand anything we must go back to a meaningful image, at a more elementary level, in a crowd **(1)**.

*Except for extremely simple forms* **(3)**, the superimposition of several images destroys each of them. We must use a more elementary level of reading, which excludes perception of the overall form of each characteristic and activates the memory.

### C.2.3. The laws of graphics

These laws correlate the problems of information processing with the properties of the image.

*Any graphic construction originates with a data table.* This table can have two objectives: providing an archive, as with a statistical yearbook, or displaying the data for a specific problem.

What are the relationships between the seasons of the year and the types of guests? This is the specific problem confronting the hotel manager.

*The bases of graphics* **183**

His goal is to discover "periods" along *x* and the "types" along *y*, that is the "global," the overall relationships of the set. And this is the information level necessary for understanding and decision-making.

1. *The aim of a graphic construction is the discovery of the groupings or orders along* x *and* y *constructed by the* z *data.*
2. *The only construction which enables us to discover these groups in every case is the* x y z *construction. It is also the simplest. It only requires:*
- *transposing the lay-out of the table, that is placing the* x *and* y *entries from the table along* x *and* y *on the graphic;*
-*transcribing the numbers in* z *by means of a size or value variation;*
- *permuting the rows and/or the columns to discover the sought-after groupings.*
3. *In the* x y z *construction, permutations and classings are the means of resolving information problems graphically.*

Two exceptions are obvious:
- *In data tables with three rows or less,* each of the rows can be represented by one of the three dimensions of the image. These are "scatterplots." They display the sought-after groupings directly.
- *Networks and topographies* denote relationships existing among the elements *of a single set*. The transcription of these relationships activates both dimensions of the plane. The *z* dimension of the image represents quantities. But a dimension is lacking for the representation of additional characteristics. Such a problem exceeds the limits of the image.

4. *Any construction with more than three characteristics, except for the* x y z *construction, or any network with more than one characteristic in* z *destroys the unity of the image and the level of overall information about the set. These are "reading" graphics.*

### C.2.4. The main problems in graphics

Graphics is a very simple language. Its laws become self-evident when we recognize that the image is transformable, that it must be reordered, and that its transformations represent a visual form of information-processing. However, three problems remain to be solved:

- *How to represent quantities in* z? This is a problem concerning steps, and it involves the *visual variables of the image (p. 186)*. Its solution depends on an implantation, by point, line, or area, of the quantities in the plane, which implies a distinction between absolute quantities and ratios. It also depends on the properties of size and value variations.

- *How to represent several characteristics in a network?* How to exceed the limits of the image while minimizing the consequences of such a decision? This is a problem concerning the visual selection of superimposed images, and it involves the *visual variables of differentiation (p. 213)*. Its solution depends on the selectivity of these variables and on the complexity of the information.

- *And above all how to construct the data table?* In brief, the laws of graphics reveal the simplicity of the graphic transcription of a data table, but this simplicity only situates the problem elsewhere. How do we portray all the data from a study or an investigation in a single table? This question does have an answer.

*In order to construct all the data of a problem as a single table we must choose one component, "objects," which we place along* x, *for which all the other components are "characteristics" that range themselves along* y. *The cells of the table must only contain quantities, ordinal numbers and/or binary answers.*

The matrix analysis presented in the final section of this book shows how to use the properties of graphics in making this major choice, which is not only a graphic problem but the problem in organizing any research and all "information-processing."


## C.2.5. The interpreting text

The presentation of any graphic processing poses the problem of word-image integration. In graphics, the image does not exist without the word, but the word has no meaning outside of the image. Here the drawing is not an illustration or a sales gimmick; it is the main part of the demonstration. Consequently, in any presentation of graphic processing, *the text must appear opposite the image:* the image on one page, the text on the opposite page. The text should be typed single-spaced. If the text

*The bases of graphics*

is more than one page, the image must appear again on the following page. If the text does not cover the entire page, the bottom part should be left blank. And one begins again with the next image.

Whenever a matrix or a file presents a relatively complex image to the reader, it is necessary to construct an interpretation matrix (pp. 57, 89, 167). Its construction obliges us to make choices, to draw lines. Remember that these choices are the very goal of the research; they become "the thesis." They must be justified in the text. With less complex images, it is advisable *to circle the groups one is citing on the image,* and to designate them by letter (p. 87). The text is simplified and integrated with the image, which makes the demonstration precise and concise.

Above all, the interpreting text should not be a simple description of the regrouped elements. This description is supplied by the interpretation matrix. With each image the text must as a general rule:

1. *Recall the hypotheses* and the classing procedure adopted.
2. *Identify the groups of characteristics* discovered. Each should be named. However, if this is impossible, say so and give an analogy with comparable groupings encountered outside the data. That is, the sociologist, the historian, the doctor . . . must learn mathematics or graphics and not, as has sometimes been claimed, the opposite.
3. *Then identify the groups of objects*, by referring to the groups of characteristics and to extrinsic analogies.
4. *Study the special cases.* They provide either an anecdote representative of the general tendency or an exception leading to new hypotheses.
5. *Requestion the "corpus"* (the set of objects) *and the characteristics* chosen by focusing on the elements which had to be eliminated in order to discover coherent groups. Remember that an investigator never says: "This is because . . . ." but only: "in this set, everything happens as if . . ."

**186**                                    *The graphic sign system*

C.3. VARIABLES OF THE IMAGE:
      THE PLANE, SIZE AND VALUE

### C.3.1. The eight visual variables

What means are at the disposal of Graphics? In order to transcribe relationships of resemblance, order and proportion, graphics utilizes eight types of variations that the eye can perceive.

A mark can, for example, appear at the top and to the left, or at the bottom and to the right on the sheet of paper. Thus it can transcribe a relationship *between the two dimensions*, x *and* y, *of the plane.*

Two marks, differently positioned on the plane, can also vary in *size, value, texture, color, orientation* or *shape*. These are the visual variables of "the third-dimension." Each of these eight visual variables can transcribe a component of the information; for example—*x*: individuals, *y*: characteristics and "size": quantities. The graphic then manifests the relationships among these three components.

The table on the facing page shows that the third-dimension variables can apply to all three "implantations"—points (P), lines (L) or areas (A) —that is to the three significations a mark can have on the plane (p. 188).

*Variables of the image*

The infinite variety of graphic constructions derives from our being able to represent any component of the information by any of the eight visual variables, provided that at least one of the two dimensions of the plane is utilized. However, the visual variables do not all have the properties: a) of manifesting an order or a proportion; and b) of being permutable.
The component manifested by the numbers in A1 on the facing page rises "in $z$" above the plane, like a "relief." This relief constructs an image in triangular form, visible only when the numbers are transcribed by size or value.
The difference between A1 and A2 is no longer visible when the other visual variables are utilized.
The image is the meaningful form that is perceived immediately.

On white paper the form of the entire set is created by a reduction in the "amount of white" at certain points on the paper. This reduction can be

*Variables of the image*

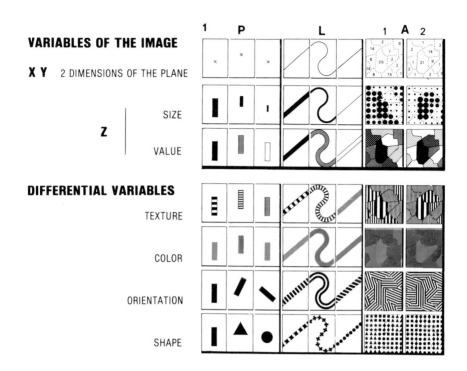

accomplished by size or value. *The variables of the image are thus the* x y *plane and the size and value variations "in z."*

*The differential variables*

The "reliefs" in A1 and A2 disappear when the transcription uses other visual variables that only underscore the flatness of the sheet of paper and the squareness of the table. Indeed, they merely differentiate the elements from each other. Any order disappears. *Texture, color, orientation and shape are differential variables.*
Accordingly, to use them to create an image of the entire set is a serious mistake.

We note, however, the exceptional property of the texture variation, which enables us to see either the two "reliefs" in $z$ or the square table.

Finally, we note that only the $x$ and $y$ variables of the plane are manipulable, that is easily permutable.

## C.3.2. The properties of the plane

*The plane manifests three basic significations.* Indeed, in **(1)**:
- A, B, C are distinguishable. They are therefore *different* ($\neq$) although they are three *similar* ($\equiv$) points.
- B is between C and A. The three points are *ordered* (O).
- BC is twice as long as AB. AB is a unit for measuring BC and AC, and the eye sees *proportions* (Q).
Consequently, the $xy$ dimensions of the plane can represent any component of the information. Any graphic construction utilizes at least one of the dimensions of the plane.

*The plane of the image is continuous and homogeneous.* We can construct an image within the image, but we cannot disregard a part of the image. Gaps in the image are fully visible. Consequently, in a signifying space:
- an absence of signs signifies an absence of phenomena and not ignorance about the phenomena,
- all visual variations appear as meaningful. A variation without meaning is thus a source of ambiguity.

## C.3.3. The plane. Implantations and the representation of quantities.

*The implantations*

In a plane, we can consider a point (P), a line (L), or an area (A).

A POINT in a plane is theoretically without surface area **(2P)**; otherwise, it would be an area. A visible mark (which must thus have area) can nonetheless signify a point, without area. This is what enables us to construct the third, $z$, dimension of the image.

A LINE is also theoretically without surface area **(2L)**, otherwise it would be an area. However, a line does have length. Consequently:
-In order to vary a line in texture, orientation or shape without changing its signification in the plane, the line is represented by a set of points, which then convey this variation (see p. 187).
- A component represented in $z$ by the width or the value of a line will be multiplied by the length of this line **(5)**. This multiplication can be a source of error.

*Variables of the image*

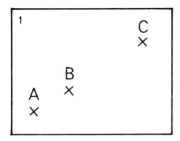
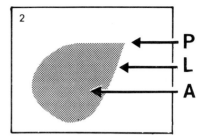
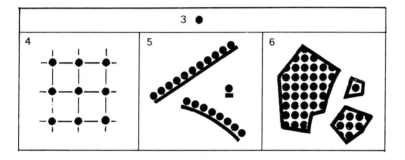

An AREA occupies a fixed portion of the plane. Consequently:
- In order to make an area vary in size, texture, orientation and shape without changing its signification in the plane, the area can be represented by a set of points or lines which manifest these variations (p. 187).
- A component represented by area in $z$ will be multiplied by the size of the area **(6)**.
As with the line, this multiplication can be a source of error; *the implantation of objects on the plane thus raises the problem of the nature of the quantities to be represented.*

Suppose we portray a population of 100 individuals by using points, lines or areas. The 100 individuals are represented by a circle **(3)**. There are indeed 100 individuals at each point of **(4)**. But there are not 100 individuals on each line of **(5)** or in each area of **(6)**. The eye sees 100 individuals multiplied by the length of the line or the size of the area, which is an obvious but nonetheless frequent mistake. What quantities can a line or an area portray?

*Graphics separates quantities into two types*
In data table **(1)**, manifesting objects A, B along *x* and characteristics P, S, P/S along *y*, the quantities are of two types: *totals per object* **(2)** and *ratios* between totals per object **(3)**. The totals are also called "absolute quantities."

*Only implantation by point can manifest absolute quantities in z.* Take a matrix **(4)** in which A, B are communities, P the quantity of population, and S the communities' surface area. The quantities are entered on points A and points B. A and B are *implanted by point*. In this case, a size variation can be used in *z* to denote the ratio P/S as well as the absolute quantities P and S **(5)**.

Take a diagram **(6)** in which A and B are represented by lengths proportional to S. A and B are *implanted by line*. In this case a variation in *y* can represent the ratio of P/S **(7)**, here 3 and 2. The surface area of the rectangles thus represents (P/S) × S = P, here 3 and 12. But if *y* represents the absolute quantity P, the rectangles then represent P × S or here 3 and 72, which is absurd **(8)**. The same is true in a map **(9)** where A and B are *implanted by area*, since, by definition, they are represented by the area of the community. Figures **(10)** and **(12)** are correct; **(11)** and **(13)** are absurd.


*Thousandths, indices, percentages, ratios, rates*
This essential difference between absolute quantities and ratios can be concealed by the following secondary transformations, which must be excluded from calculations and representations.
For *thousandths* (or hundredths) multiply all the numbers in a series of absolute quantities by a constant: 1000/total of the series. These are absolute quantities.
For *indexing indices* multiply all the numbers in a series of absolute quantities by a constant: 100/B, B being a base value. These are absolute quantities.
For *percentages* multiply all the numbers in a series of *proportions* by a constant: 100. These are ratios.
It must be understood that these three transformations change nothing in the two basic categories: absolute quantity per object and the ratio between two absolute quantities. These transformations change only the *verbal* scale of the numbers. We may establish that these numerical transformations do not transform the graphic image.

*Variables of the image*

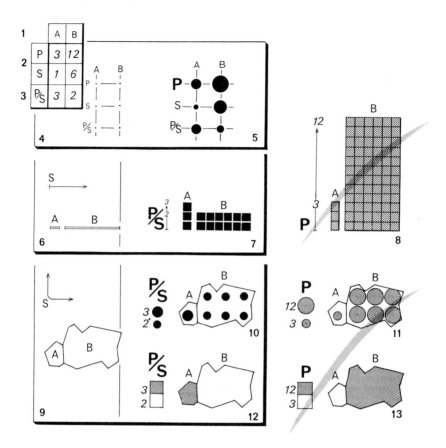

*Rates, like ratios,* are proportions. The word "index" may also designate a proportion. This is an "index-indicator." It must not be confused with the indexing index, which is an absolute quantity.

To avoid any confusion that may result from the use of this complicated terminology, one need only return to the origin of the numbers. How are they calculated? For a ratio, the correct designation is simply the expression of the two terms of the ratio. For thousandths, hundredths, percentages or indices we must be able to answer the question: "for one hundred what?" We will confirm that this answer is sometimes quite difficult to obtain.

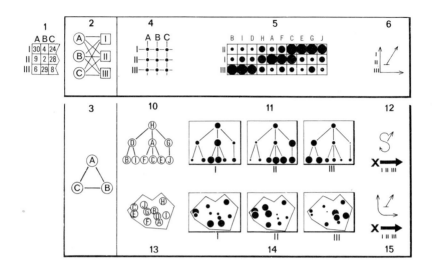

## C.3.4. Impositions on the plane: diagrams, networks and topographies

Information establishing relationships between two sets constructs a *diagram* **(2)**. Information establishing relationships among the elements of a single set constructs a *network* **(3)**. A network ordered according to an arrangement of elements in an object is a *topography* **(13)**.

*Diagram constructions*

To transcribe table **(1)**, we can either portray *elements by points* and relationships by lines **(2)** or, conversely, represent *elements by lines* and relationships by points **(4)**. The classic construction **(2)** displays the difference between information **(2)** and **(3)**, but we must recognize that its further development is useless. One must look for the aberrant relationship in **(7)**. It appears immediately in **(8)**. *Only construction* **(4)** *can be further developed* without destroying legibility.

But why not one of the constructions in **(9)**, that follow the same principle? Simply because a construction must only depict meaningful visual variations. However, any visual length other than a straight line, any variation of distance between straight lines, any non-right angle is not

*Variables of the image* 193

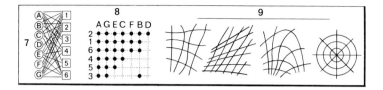

neutral and creates meaningless variations. *Only construction* **(4)** provides neutrality, that is the "all things being equal" of graphic transcription. These two observations explain the universality of the "double entry table." This construction can be schematized as in **(6)**. Other constructions are imaginable, but their uselessness will be illustrated on the following page.

*Network constructions.* Take **(1)** cities A, B, C... related to characteristics I, II, III. These data can construct diagram **(5)**, a matrix construction (objects along $x$, characteristics along $y$). But add *new information*: the hierarchy of administrative relationships **(10)** among these cities. Its transcription mobilizes $x$ and $y$. The quantities are transcribed in $z$ and . . . a dimension is lacking to represent the different characteristics! Therefore, it requires *at least* as many moments of perception as there are characteristics **(11)** to see the entire set of relationships. This construction is schematized by **(12)**. (S) represents the pattern of points, multiplied by the number of characteristics, one image per characteristic.

Networks are by definition reorderable. The main problem with networks involves their simplification through transformation of the pattern, in order to eliminate meaningless intersections (p. 131).

*Topographies are ordered networks* **(13)**. Accordingly, they do not pose any construction problem on the plane. The notion of order depends on the constancy of the physical object they represent: an anatomy, a cell, a machine, a land section, geography, cosmography . . . Topographies draw all their properties from this constancy (see p. 139).

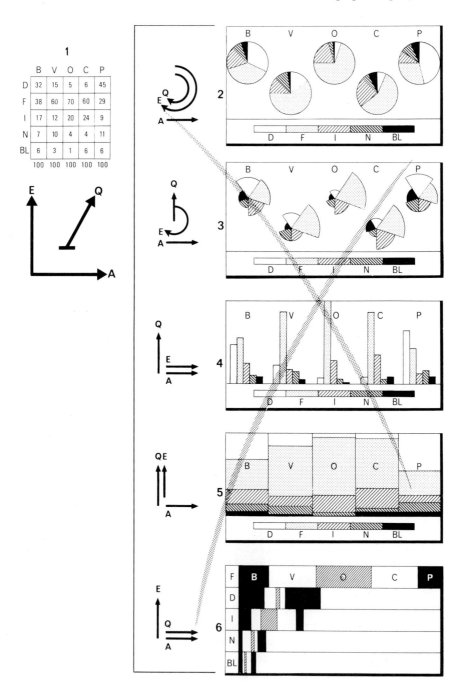

## Variables of the image

*Useless diagrams.*
Consider data **(1)** indicating the distribution of meat production in the Common Market countries in 1966\*. Are the ministers who are studying the meat market really aided by having figure **(2)** before their eyes? They can only pull out a few ratios; table **(1)** is much more efficacious. How are the five nations grouped or juxtaposed within the context of this information on production? None of the constructions **(2)** to **(6)** provides an answer. These diagrams are useless.

The standard construction **(8)**, however, displays information about the entire set and reveals a basically political problem: the nations form two groups—Germany and Holland on the one hand, Italy and France on the other—that can only be politically at odds or aligned. And in this case, Belgium-Luxemburg has the role of arbiter.
The standard construction does not modify the structure of the data table. It applies the $x$ and $y$ components of the table to the $x$ and $y$ axes of the sheet of paper **(7)**. Any other construction reduces the level of perceived information.

The standard construction entails the elimination of useless differences and consequently the bringing together of similar rows and, where possible, similar columns. It entails a "transformation" of the data matrix. Image **(8)** is not in the same order as the data table **(1)**. The necessity of these transformations, that is permutations, can be schematized as in **(9)**.

\*B. Beef, V. Veal, O. Lamb, C. Horse, P. Pork. D. Germany, F. France, I. Italy, N. Holland, BL. Belgium-Luxemburg.

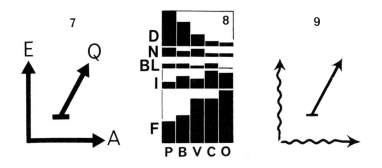

### C.3.5. The Z dimension: properties of size and value

The third or *z* dimension of the image represents a reduction in the amount of white for each element of the plane. It is obtained by variations in size or value.

*Size is "quantitative"* (Q). A difference in size can portray a proportion between two magnitudes: A is twice as large as B. B is the unit which serves to measure A.

White cannot serve as a unit for measuring black. Consequently, *value is not quantitative.*

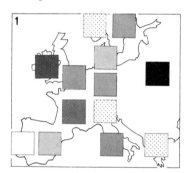
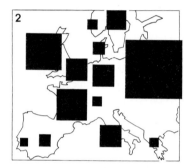
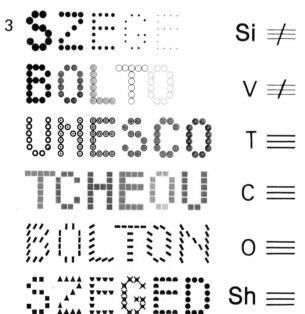

## Variables of the image

For example, it cannot portray consumption in the different countries in **(1)**. What is consumption in Italy if we know that Portugal consumes **1**? A *size* variation **(2)** enables us to make a correct approximation.

*Size and value are ordered* (O). There is no need for a legend when putting sizes or values into order. They create an immediate order which obviously must correspond to the order of the component. To not make the visual order correspond with the order provided by the data would be a grievous irresponsibility on the part of both researcher and designer. In an ordered transcription the legend serves only to "verbally" define the limits of the steps.

*Size and value are dissociative* ($\neq$): they have variable visibility: the counterpart of their essential function: to create images. It is not possible to exclude the variation in visibility: it dissociates the names in **(3)**, whereas the other visual variables have constant visibility and are said to be associative ($\equiv$).
Consequently, size and value dissociate any other variable with which they are combined. For example, the number of distinguishable colors decreases with the size and/or value of the colored marks.

*The representation of quantities in* z poses two problems:
- the problem of implantation: absolute quantities are only representable in implantation by point (p. 190);
- the problem of steps, that is, the correspondence between the series of numbers and a series of values or sizes (p. 198).

### C.3.6. The problem of steps in Z

C.3.6.1. This is one of the most interesting and most frequently discussed problems in graphics, since it has no intrinsic solution. Indeed, its solution must rely upon information which is extrinsic to the data.

When one of the two dimensions of the plane is available to portray quantities **(1**, p. 199), the problem does not arise. It must be dealt with, however, when neither $x$ nor $y$ is available, that is, in two cases:

a) When the area for portrayal is very small, as in (2). Remember that this reduction in size is the price which must be paid for the portrayal of *n* characteristics in a matrix (3) or of quantities on a map (6) and (7).
b) When we use value variations (4) and (5).
In both cases it is a matter of MAKING THE PERTINENT GROUPINGS VISIBLE, that is those constructed by the numbers, as in (1). The problem differs according to whether we utilize size or value.

*Difference of perception between size and value*
When the elementary marks are very small, they are perceived in groups (3, 4 and 5). We perceive VALUES. A "grey" has a certain proportion of "black" and of "white." This is a ratio: for example, 10% black and 90% white or 10/90.
The differences between two values is therefore a ratio of ratios, for example $\frac{10}{90}$ / $\frac{20}{80}$. When these "ratios of ratios" conform to the general law of perception (constant proportionality), we obtain a sequence of "equidistant"* values (9). A linear distribution of quantities (8) will be denoted in *z* by a series of equidistant values (9).

*When the quantities are transcribed by values, perception is based on the visual equidistance of the value steps. Consequently, the corresponding quantities have an arithmetic progression* (10), for example, 1, 2, 3, 4, 5 . . .

Any other division creates erroneous groupings. With legend (11), for example, the straight line (12) will be perceived as a curve (13).

When the size of the elementary mark enables us to see a difference between two signs (6 and 7), we perceive a size difference independent of the amount of white surrounding the signs.
The difference between two sizes is therefore a simple ratio. When this ratio conforms to the general law of perception, constant proportionality, the surface area of the signs has a geometric progression **(14).**

*When the quantities are transcribed by sizes, perception is based on the geometric progression of the surface area of the signs. Consequently, the corresponding quantities have a logarithmic progression* (15), for example, 1, 2, 4, 8, 16 . . .
Any other division creates erroneous groupings.

*This equidistance is defined by the equal visibility of the lines created by the difference between two successive values (9) (see *S.G.*, p. 75).

## Variables of the image

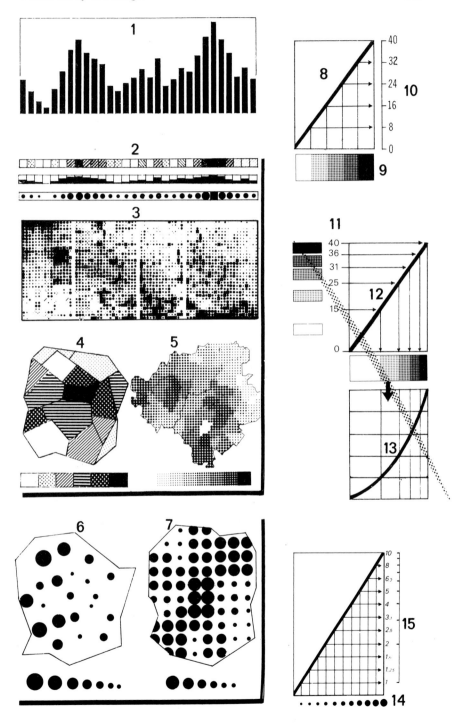

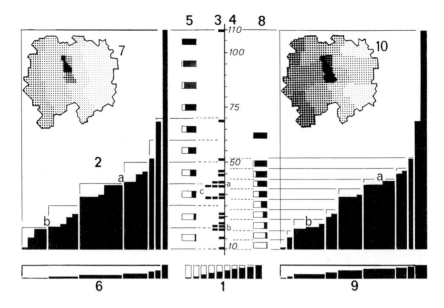

## C.3.6.2. The Z Dimension: equidistant degrees of value

Take a series of quantities and a scale of ten equidistant values or ten signs **(1)** which will be perceived as values because of their small size or their groupings, as in a matrix.

If we can use the *y* dimension of the plane to represent the quantities, we can construct a repartition **(2)** or a distribution **(3)**. They display the groupings which must not be destroyed. It is now a matter of establishing a correspondence between the scale of quantities **(4)** and the series of signs available in *z* **(1)**.

Let us consider the correspondence in **(5)**. It produces a row **(6)** and a map **(7)**, both almost white. It does not fulfill the first aim of the operation: making the groupings VISIBLE. If we look closely at the groupings it is supposed to define, we observe that it breaks up groups *a* and *b* and

## Variables of the image

constructs a false group *c*. Therefore, neither does it fulfill the second aim of the operation: making the PERTINENT GROUPINGS visible.

On the other hand the correspondence in **(8)** does make the pertinent groupings, *a* and *b*, visible. What is the difference between **(5)** and **(8)**?

*The shifting of bounds*
In **(5)**, the step series in $z$ is extended out to the extremities of the series of quantities. The "bounds" are 10 and 110. Consequently this leaves only a small visual distance: three steps to represent most of the numbers in the series. As a result, the groupings are barely visible and are not pertinent. In **(8)**, the bounds are 10 and 52. The two largest numbers are included in the highest step of the series. As a result, the entire distance between black and white remains available for representing most of the numbers. Row **(9)** and map **(10)** do display the pertinent groupings.

*This is a problem of visual distance.* The visual distance between black and white within the limited areas of a transcription in $z$ cannot even be compared with the visual distance available on a plane. One need only look at **(2)** and **(6)**. It is thus necessary to adjust the total available distance. This is not a problem of the number of steps. We could imagine using 20, 50, or 500 steps between white and black, but row **(6)** would not be left any less white, and even more pertinently the steps would have nevertheless remained invisible! An increase in the number of steps in no way increases the total distance available and does not solve the problem.

*The choice of bounds*
The choice in **(8)** places the two largest numbers (the "upper extremes") in the last step of the series, and it is debatable. It would not be illogical to choose 10 and 70 instead of 10 and 52, or to consider the quantity 10 as a "lower extreme"* and to choose 15 and 52 or 15 and 70.

Let us remember the two imperatives which direct this choice.

VISIBILITY: the row of the matrix or map must be neither too white nor

---

*These "statistical extremes" can be marked in matrices and maps, by using a black point in the white areas or a white point in the black areas.

too black; otherwise the groupings will be invisible and the representation useless. PERTINENCY: the groupings constructed by the numbers must not be destroyed.

Rationalizations of this choice, based on the mean, the median or the standard deviation, do not ensure the fulfillment of these two imperatives. To satisfy them, it is necessary to study the distribution of the numbers and the nature of the extremes.

*Practical construction of a step scale*

*a)*-Construct the distribution of the quantities. Using graph paper, place the smallest and largest numbers at the extremities of a suitable, sufficiently large scale: about 15 to 30 cm.

-Successively plot each number at its place on the scale. The points defined by the same number are superimposed **(1)**. For a large quantity of numbers, divide the scale into classes. The number of classes should be several times larger than the number of available steps **(2)**.

*b*-Study the distance between and the nature of the extremes. They often define objects of a different nature in relation to the characteristic being studied. The population density in urban areas is not of the same nature as the density of other regions. The vineyard area in non wine-producing regions is not of the same nature as that in wine-producing regions. Starting from this fact, we can group several extreme numbers, even though there might still be large distances between them in the number scale.

*c*-Vary the spacing and the position of the bounds. For this operation, we can construct on a separate piece of paper a very simple graph **(3)**, corresponding to the number of available steps and on which we can shift the distribution **(4)**.

Study the correspondence between the groupings visible on the distribution and the step limits. A slight shift may suffice to either break up a grouping or, on the contrary, to define it.

When the scale is determined, it is sufficient to read from the graph paper the numbers which define the limits of each step.

It is normal for certain steps not to be represented. Thus the visual distance translates a distance between groupings. It is not illogical, when the groups are clearly separated, to associate the same step with all the elements of the group by slightly extending the step's limits. In certain cases, we can slightly reorient the graph **(5)**, thereby progressively increasing distances and enabling us to take the particularities of certain characteristics into account.

*Variables of the image* 203

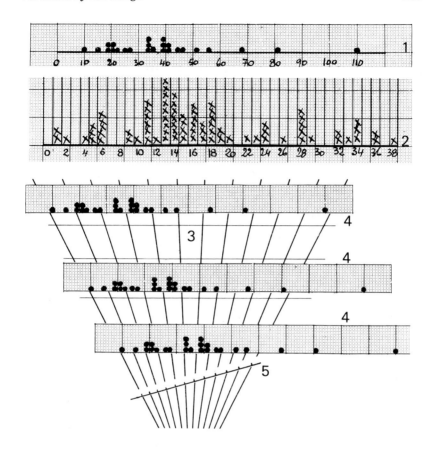

***Should the mean be in the central step?*** This question raises another one: are we more interested in representing the groupings or the mean? When it is a matter of *comparing numerous characteristics*, it is important to define the groupings as strictly as possible. When it is a matter of *discussing a single characteristic,* it is convenient to define the objects in relation to the mean object. However, this does not imply that it must be represented by the central step (see p. 65).

The graphic sign system

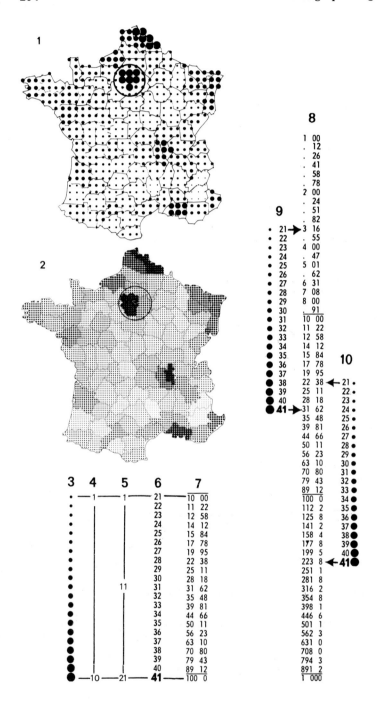

*Variables of the image*

## C.3.6.3. The Z Dimension: proportional sizes

The representation of quantities by proportional sizes is only applicable when the signs remain sufficiently large **(1)**. A considerable reduction of the signs leads to the perception of values **(2)** and consequently to the law of equidistance (p. 198).

*A series of proportional circles*

**(3)**-Consider a series of circles.
**(4)**-The surface area of the largest is ten times that of the smallest.
**(5)**-Between these extremes, twenty steps provide a slight but still perceptible difference between two successive circles. With a larger number of steps, the difference would be invisible and therefore useless. Consequently, this series represents the maximum number of usable steps accepting a decimal division.
**(6)**-These circles are numbered from 21 to 41.
**(7)**-The area of a circle is equal to that of the preceding one multiplied by 1.122. The ratio between two successive circles is therefore constant (logarithmic progression).


Consequently this series of circles possesses the following properties:
- It conforms to the general law of perception: we only perceive ratios.
- The progression can be extended at either end **(8)**.
- A ratio is independent of the numbers expressing it. In the scale of surface areas, circle number 21 can correspond to any number. Circle 41 will always represent a number ten times larger **(9)** and **(10)**, and the intermediate numbers will correspond only to the perceptible steps.
- The correlation of the series of circles with an arithmetic series: 10, 20, 30, 40 etc. . . . destroys the quantitative analogy. In order to transcribe such a series by corresponding surface areas, it would be necessary, starting with circle 21, for example, to choose circles 21, 27, 31, 33 . . . .

*Utilization of the series in implantation by point*

The representation of quantities by corresponding surface areas is the least bad quantitative analogy that we can construct in $z$. Consider a series of cities in which the population runs from 16,000 to 140,000 inhabitants. To represent these numbers by surface areas, it is sufficient to

# The graphic sign system

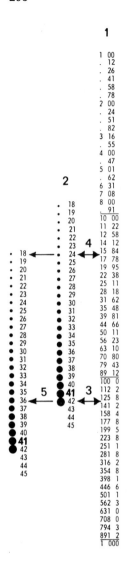

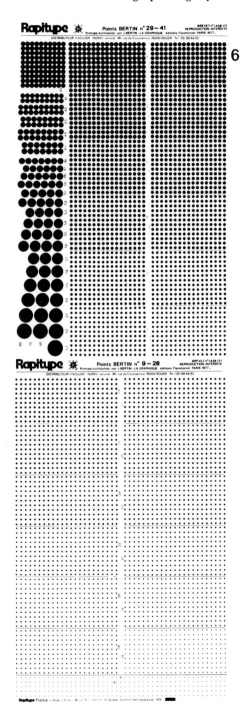

## Variables of the image

choose the size of the circle which will represent the largest population. If circle number 42 is suitable:
- Cut out from the chart (p. 208) the rule with the numbers which label the circles **(2)**.
- Place number 42 up against the scale of circle areas **(1)**, opposite the interval which contains 140,000, that is, between 125.8 and 141.2 **(3)**. All the cities included between 125,800 and 140,000 inhabitants will be represented by circle number 42. A city with 16,000 inhabitants is represented by circle number 24 **(4)**. For any population there will be a circle of proportional surface area. For example, a city with 44,600 inhabitants will be represented by circle number 32.
- If circle number 42 is too large, it is sufficient to move the rule and choose number 36, for example, as the largest circle **(5)**. The smallest city (16,000) is then represented by number 18. The areas are smaller, but proportions are preserved. The size of the largest circle is a matter of legibility. Avoid secant circles. A completely out-of-scale circle can be made "transparent" (**1** and **2** p. 204).

### Utilization of the series in implantation by area

Consider quantities applied to areas, say communities, for example.
- These quantities must be ratios (p. 190).
- To extend these quantities over the entire area, we must apply a grid of points on the map. The scale of this grid must be such that there is at least one point in each area.
- The correspondence between quantities and proportional circles is determined as in implantation by point.

However, the series of circles is limited toward the bottom. Circle number 1 is 0.5mm in diameter, and the variation between the smallest circles goes below the threshold of differentiation. The series thus preserves only the perceptible differences, that is circles number 1, 9, 15, 18, 19 . . .

But in implantation by area the series is also limited towards the top by the spacing of the pattern: 5mm in prefabricated plates **(6)**. Circle number 41 corresponds to tangent circles.*

Consequently, the representation of areal data by circles of proportional size is possible whenever the ratio between the extremes is at least 1/10 (otherwise the image risks being poorly differentiated) and at most 1/100. However, we should note that circle number 1 is minute and often disappears when reproduced. It is not advisable to use it. This reduces

*Plates developed by the Laboratoire de Graphique, Ecole des Hautes Etudes en Sciences Sociales, 131, bd St. Michel 75005, PARIS.

# The graphic sign system

| A | B | C | D | E | F | G |
|---|---|---|---|---|---|---|
| $S=\sqrt{Q}$ | $S=Q^{\frac{2}{3}}$ | $S=Q$ | $S=Q^2$ | $S=Q^3$ | $S=Q^4$ | $S=Q^6$ |

| | A | B | C | D | E | F | G |
|--|--|--|--|--|--|--|--|
| 1 | 1 00 | 1 00 | 1 00 | 1 00 | 1 00 | 1 000 | 014 • 1 |
|  | . 26 | . 19 | . 12 | . 090 | . 06 | . 030 | 044 |
|  | . 60 | . 41 | . 26 | . 188 | . 12 | . 059 | 074 |
|  | 2 00 | . 70 | . 41 | . 292 | . 19 | . 090 | 106 |
|  | . 50 | 2 00 | . 58 | . 412 | . 26 | . 122 | 138 |
|  | 3 16 | . 38 | . 78 | . 540 | . 33 | . 155 | 171 |
|  | 4 00 | . 82 | 2 00 | . 678 | . 41 | . 188 | 205 |
|  | 5 00 | 3 35 | . 24 | . 830 | . 50 | . 223 | 241 |
| 9 | 6 31 | 4 00 | . 51 | . 995 | . 58 | . 258 | 277 • 15 |
|  | 8 00 | . 73 | . 82 | 2 175 | . 68 | . 292 | 314 • 18 |
|  | 10 0 | 5 62 | 3 16 | . 371 | . 78 | . 333 | 353 • 20 |
|  | 12 58 | 6 63 | . 55 | . 585 | . 88 | . 372 | 392 • 22 |
|  | 15 84 | 7 94 | 4 00 | . 818 | 2 00 | . 412 | 433 • 24 |
|  | 19 95 | 9 44 | . 47 | 3 072 | . 11 | . 453 | 475 • 26 |
| 15 | 25 11 | 11 22 | 5 01 | . 349 | . 24 | . 496 | 518 • 28 |
|  | 31 62 | 13 33 | . 62 | . 652 | . 37 | . 540 | 562 • 30 |
|  | 39 81 | 15 84 | 6 31 | . 981 | . 50 | . 584 | 608 • 32 |
| 18 | 50 11 | 18 83 | 7 08 | 4 340 | . 66 | . 631 | 654 • 34 |
| 19 | 63 10 | 22 38 | 8 00 | . 731 | . 82 | . 678 | 703 • 36 |
| 20 | 79 43 | 26 60 | . 91 | 5 158 | 3 00 | . 727 | 753 • 38 |
| 21 | 100 | 31 62 | 10 00 | . 623 | . 16 | . 778 | 804 • 40 |
| 22 | 125 8 | 37 58 | 11 22 | 6 130 | . 35 | . 830 | 856 • 42 |
| 23 | 158 4 | 44 66 | 12 58 | . 633 | . 55 | . 883 | 911 • 44 |
| 24 | 199 5 | 53 08 | 14 12 | 7 286 | . 76 | . 938 | 966 • 46 |
| 25 | 251 1 | 63 10 | 15 84 | . 943 | 4 00 | . 995 | 2024 • 48 |
| 26 | 316 2 | 74 98 | 17 78 | 8 660 | . 22 | 2 053 | 083 • 50 |
| 27 | 398 1 | 89 12 | 19 95 | 9 441 | . 47 | . 113 | 144 • 52 |
| 28 | 501 1 | 104 4 | 22 38 | 10 03 | . 73 | . 175 | 206 • 54 |
| 29 | 631 0 | 125 8 | 25 11 | 11 22 | 5 00 | . 238 | 271 • 56 |
| 30 | 794 3 | 149 6 | 28 18 | 12 23 | . 31 | . 304 | 337 • 58 |
| 31 | 1 000 | 177 8 | 31 62 | 13 33 | . 62 | . 371 | 405 • 60 |
| 32 | 1 258 | 211 3 | 35 48 | 14 53 | . 96 | . 441 | 476 • 62 |
| 33 | 1 584 | 251 1 | 39 81 | 15 84 | 6 31 | . 511 | 548 • 64 |
| 34 | 1 995 | 298 5 | 44 66 | 17 27 | . 63 | . 585 | 622 • 66 |
| 35 | 2 511 | 354 8 | 50 11 | 18 83 | 7 08 | . 660 | 699 • 68 |
| 36 | 3 162 | 421 6 | 56 23 | 20 53 | . 50 | . 738 | 778 • 70 |
| 37 | 3 981 | 501 1 | 63 10 | 22 38 | . 94 | . 818 | 859 • 72 |
| 38 | 5 011 | 595 6 | 70 80 | 24 41 | 8 42 | . 901 | 943 • 74 |
| 39 | 6 310 | 708 0 | 79 43 | 26 60 | . 91 | . 985 | 3 028 • 76 |
| 40 | 7 943 | 842 2 | 89 12 | 29 01 | 9 44 | 3 072 | 117 • 78 |
| 41 | 10 000 | 1 000 | 100 0 | 31 62 | 10 00 | . 162 | 208 • 80 |
| 42 | 12 580 | 1 188 | 112 2 | 34 47 | 10 59 | . 255 | 302 |
| 43 | 15 840 | 1 412 | 125 8 | 37 58 | 11 22 | . 349 | 413 |
| 44 | 19 950 | 1 678 | 141 2 | 40 97 | 11 88 | . 447 | 502 |
| 45 | 25 110 | 1 995 | 158 4 | 44 66 | 12 58 | . 548 | 594 |
| 46 | 31 620 | 2 371 | 177 8 | 48 70 | 13 33 | . 652 | 689 |
| 47 | 39 810 | 2 818 | 199 5 | 53 08 | 14 12 | . 758 | 813 |
| 48 | 50 110 | 3 349 | 223 8 | 57 87 | 14 96 | . 868 | 924 |
| 49 | 63 100 | 3 981 | 251 1 | 63 10 | 15 84 | . 981 | 4 039 |
| 50 | 79 430 | 4 731 | 281 8 | 68 78 | 16 78 | 4 097 | 157 |
| 51 | 100 000 | 5 623 | 316 2 | 74 98 | 17 78 | . 216 | 297 |
| 52 | 125 800 | 6 663 | 354 8 | 81 75 | 18 83 | . 340 | 403 |
| 53 | 158 400 | 7 943 | 398 1 | 89 12 | 19 95 | . 466 | 525 |
| 54 | 199 500 | 9 441 | 446 6 | 97 16 | 21 13 | . 597 | 664 |
| 55 | 251 100 | 11 220 | 501 1 | 105 9 | 22 38 | . 731 | 800 |
| 56 | 316 200 | 13 330 | 562 3 | 115 5 | 23 71 | . 870 | 940 |
| 57 | 398 100 | 15 840 | 631 0 | 125 8 | 25 11 | 5 011 | 084 • |
| 58 | 501 100 | 18 830 | 708 0 | 137 2 | 26 60 | . 158 | 233 |
| 59 | 631 000 | 22 380 | 794 3 | 149 6 | 28 18 | . 308 | 386 |
| 60 | 794 300 | 26 600 | 891 2 | 163 1 | 29 85 | . 463 | 543 |
| 61 | 1 M | 31 620 | 1 000 | 177 8 | 31 62 | . 623 | 705 |
| 62 | 1 258 | 37 580 | 1 122 | 193 8 | 33 49 | . 787 | 871 |
| 63 | 1 584 | 44 660 | 1 258 | 211 3 | 35 48 | . 956 | 6 043 |
| 64 | 1 995 | 53 080 | 1 412 | 230 4 | 37 58 | 6 130 | 219 |
| 65 | 2 511 | 63 100 | 1 584 | 251 1 | 39 81 | . 310 | 401 |
| 66 | 3 162 | 74 980 | 1 778 | 273 8 | 42 16 | . 494 | 587 |
| 67 | 3 981 | 89 120 | 1 995 | 298 5 | 44 66 | . 633 | 780 |
| 68 | 5 011 | 104 400 | 2 238 | 325 5 | 47 31 | . 878 | 978 |
| 69 | 6 310 | 125 800 | 2 511 | 354 8 | 50 11 | 7 080 | 182 • |
| 70 | 7 943 | 149 600 | 2 818 | 386 8 | 53 08 | . 286 | 392 |
| 71 | 10 M | 177 800 | 3 162 | 421 6 | 56 23 | . 498 | 608 |
| 72 | 11 220 | 211 300 | 3 548 | 459 7 | 59 56 | . 718 | 830 |
| 73 | 15 840 | 251 100 | 3 981 | 501 1 | 63 10 | . 943 | 8 058 |
| 74 | 19 950 | 298 500 | 4 466 | 546 3 | 66 33 | 8 175 | 293 |
| 75 | 25 110 | 354 800 | 5 011 | 595 6 | 70 80 | . 424 | 536 |
| 76 | 31 620 | 421 600 | 5 623 | 649 4 | 74 98 | . 660 | 785 |
| 77 | 39 810 | 501 100 | 6 310 | 708 0 | 79 43 | . 912 | 9 042 |
| 78 | 50 110 | 595 600 | 7 080 | 771 8 | 84 24 | 9 173 | 305 |
| 79 | 63 100 | 708 000 | 7 943 | 842 4 | 89 12 | . 441 | 587 |
| 80 | 79 430 | 842 400 | 8 912 | 917 3 | 94 41 | . 716 | 856 |
|  | 100 M | 1 M | 10 000 | 1 000 | 100 | 10 |  |

J. BERTIN

*Variables of the image* 209

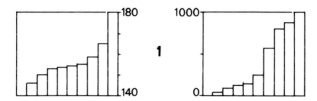

the available ratio to 1/70, taking into account the necessity of complete black above circle number 41.

For higher ratios, it is necessary either to eliminate statistical extremes (p. 201) or to use a different scale than that of surface areas, or both.

*Extensive scales*

How can we represent quantities when the ratio between the extreme numbers is smaller than 1/10 and the bounds are 20 and 50, as with output, or 140 and 180, as with the height of individuals?

Conversely, how can we represent quantities with a very large range, from 1 to 1000, for example, as with certain densities? In diagrams, when the *y* dimension represents the quantities, we adapt the quantities to the nearly constant* height of a diagram (**1**). This adaptation is accomplished by the scales on the facing page.

Scale C is the scale of the circles' surface areas. Between circle number 1 and circle number 41, the ratio is 1/100. The chart gives the scales for the following ratios between numbers 1 and 41:

A: 1/10,000, B: 1/1,000, C: 1/100, D: 1/31, E: 1/10, F: 1/3.

Finally, if we reduce by half the scale of the numbers of points (G), we can read on F a ratio of 1/1.7 between number 1 and number 41.

After having chosen the scale corresponding to the range of the series to be represented, it suffices, as before, to place number 41 on the rule opposite the largest number, in order to directly read the numbers, that is the steps corresponding to the circles of the series. *It is this series that determines class intervals.* To work inversely is the same as poorly defining the problem, discovering non-existent groups, and implementing incorrect decisions.

---

*A graphic too high or wide disrupts the balance of *x* in relation to *y* and becomes illegible. A graphic which is too large or too small is inconvenient. This "constancy" is thus defined by the properties of visual perception.

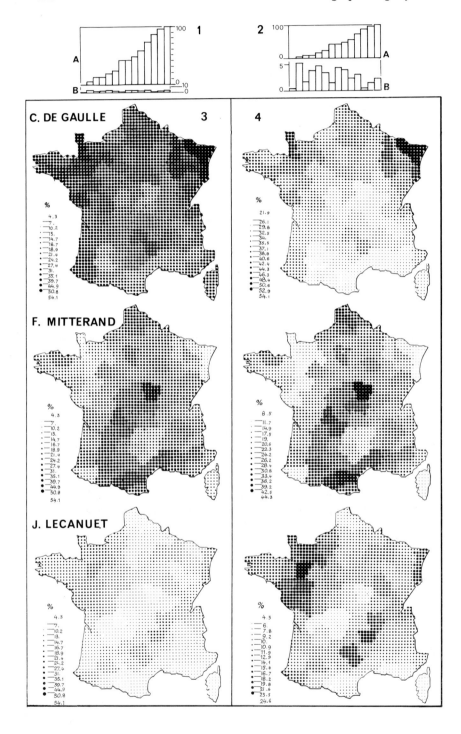

*Variables of the image* 211

## C.3.6.4 Common scales—Proper scales

Should several rows in a diagram or matrix or several maps in a collection have a common scale for the quantities in $z$ or, on the contrary, should each have its own individual scale or "proper" scale? Transposed onto $y$ on the plane, this question can be formulated as follows: should we construct **(1)** or **(2)**?

This is a question of visibility. Should we display the pertinent groupings constructed by the components, or should we sacrifice the groupings for the benefit of the perception of the totals?

In **(1)** the totals are immediately visible, but the groupings and distribution of B completely invisible. In **(2)** the contrast between A and B is immediate, and the totals can be expressed by a number, a sign or by the scale.

*In diagrams* a common scale is often possible, provided that the profile of the smallest rows is perfectly visible and comparable to the other profiles.

*In automated statistical cartography,* we can use both scales. The scale is *common* for maps **(3)** and runs from the highest percentage obtained, 54% by C. DeGaulle, to the lowest, 4% by J. Lecanuet. But the local influence of J. Lecanuet is not readily apparent. The three maps stress the overall results that three numbers could sufficiently convey. In **(4)** each map has its *proper* scale using the whole range of visual perception. The characteristic geographic distribution for each candidate is then apparent (in terms of percentage of voters registered for the 1965 election).

*In matrices and image-files* the proper scale is the general rule, since it is in these constructions that the available visual variation is smallest. But very homogeneous situations sometimes authorize a common scale.

Take, for example, the transcription of the number of days of vacation in the matrix on page 44. This number ranges from 10 to 28 and is relatively homogeneous across the four professions. *The common scale is adopted.* It clearly describes each distribution and thus each "geography." Moreover, it allows us to compare professions, among themselves, and to discover that independent of geography the cashier and the bus driver generally have more vacation time than the secretary or mechanic.

In this same matrix, "free time" is the complement of the number of working hours. It is very comparable across four of the professions. A common scale is therefore adopted. But the "free time" attributed to the school teacher is generally much greater. A proper scale for the profession of school teacher saves us from constructing an almost entirely black row.

*The graphic sign system*

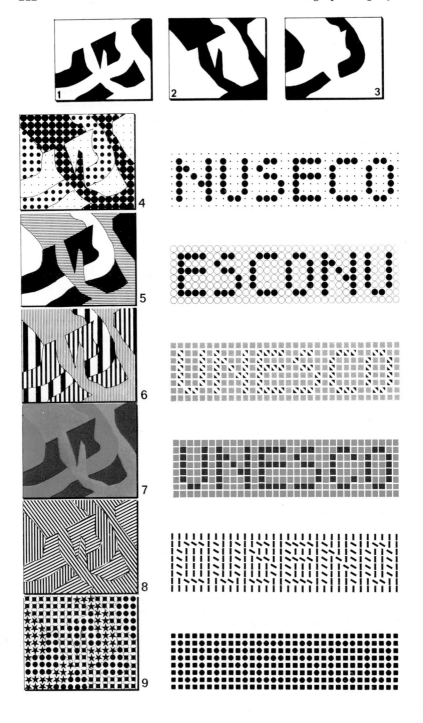

## C.4. DIFFERENTIAL VARIABLES

### C.4.1. What does it mean to distinguish images?

Image **(4)** separates or "differentiates" images **(1)**, **(2)** and **(3)** by a size variation. This differentiation is accomplished by value in **(5)**, texture in **(6)**, color in **(7)**, orientation in **(8)**, and shape in **(9)**.

*To distinquish a characteristic* is, for example, to see in **(4)** all the elements of image **(1)** and thereby *visually disregard the rest*. This involves reading the word in images **(10)**. We can see that, all things being equal, the third dimension variables do not all have the same differential properties. Differentiation depends on the visual variable used. But in actuality it is also inversely related to the number of characteristics and to the complexity of their distribution on the plane.

*Size* **(4)** *and value* **(5)** provide an excellent differentiation among different characteristics ($\neq$). They create an image. But they also construct a visual hierarchy which favors a particular characteristic, for example, characteristic **(2)** in image **(4)** or characteristic **(1)** in image **(5)**. *This hierarchy is an error* when unjustified, as, for example, in the density maps on page 170. *When the characteristics are quantitative*, size and/or value are used to represent the quantities (or their order). They are no longer available to differentiate among characteristics. Take, for example, the four images in **(2)**, p. 182. To distinguish among them, in figure **(1)** on p. 182, we must call upon differential variables.

*The differential variables:* texture, color, orientation, and shape give the same visibility to each characteristic; they are associative ($\equiv$). They can thus combine with size and/or value without modifying their properties. The following three observations are noteworthy:
- Absolute selectivity, independent of the number of characteristics or of their distribution, only exists for differentiation on the plane **(1)**, **(2)** and **(3)**.
- Color **(7)** is the variable which provides the least bad selectivity, after size and value (under normal lighting).
- Shape **(9)** has no selectivity. To distinguish one shape (at the higher level) constructed by shapes (at the elementary level) obviously requires as many instants of perception as there are elementary shapes. But ideal selectivity would mean the reduction of the number of instants to one, as in figure **(1)** for example.

# The graphic sign system

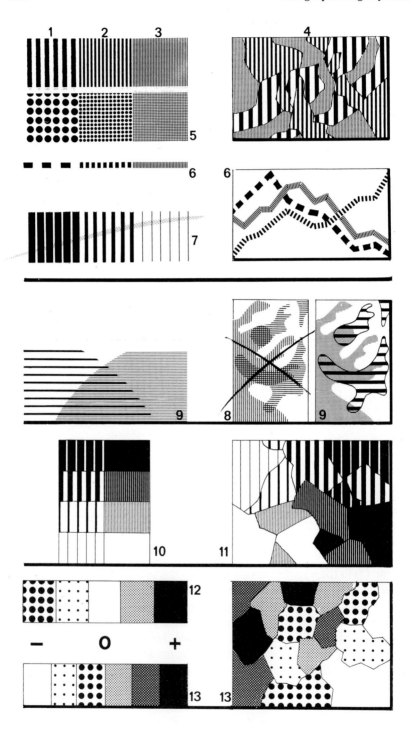

*Differential variables* 215

## C.4.2. Variation by texture

This involves the "photographic" reduction of a given pattern, as with the difference among **(1)**, **(2)** and **(3)** in a line pattern or the difference among the three steps of **(5)** in a point pattern. Variation by texture is independent of the shapes in the pattern.* We must not confuse variation by texture **(1, 2, 3)** which does not vary the visibility of the steps, with size variation **(7)** which does vary their visibility.

*Texture is selective* (#). It enables us to distinguish among areas **(4)** or lines **(6)** of the same visibility. It can provide three steps of selectivity. Texture is "transparent" and enables us to differentiate superimposed areas **(9)**. It avoids confusion **(8)**.

*Texture is associative* ( ≡ ). Having constant visibility, it can be associated with value in order to differentiate *two scales of value* **(10)**, which enables us to see in **(11)** a variation ordered from left to right combined with a vertical difference of characteristics (see p. 83).
It can also separate *a single scale of values into two*, for example, positive and negative as in **(12)**, or preferably, as in **(13)**. In problems **(6, 9, 11** and **13)** variation by texture is generally the most efficacious graphic solution.

*Texture is ordered* (O). It has the exceptional property of creating an order among characteristics without downplaying any of them. The sum of the images remains visible. In **(4)**, for example, the areas are all equally visible, though nonetheless ordered by texture.
Finally, the notion of texture is fundamental in problems of reproduction. Numerous prefabricated screens have a very fine texture that disappears in reductions and reproductions. Printers refuse them, and we have to re-do the drawing.

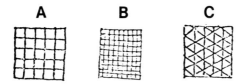

*A and B differ in their texture, but there is no difference in pattern. The elementary shapes are the same. The notion of pattern explains the difference between A and C. The elementary shapes are different. A difference in "pattern" is essentially a difference in shape.

# The graphic sign system

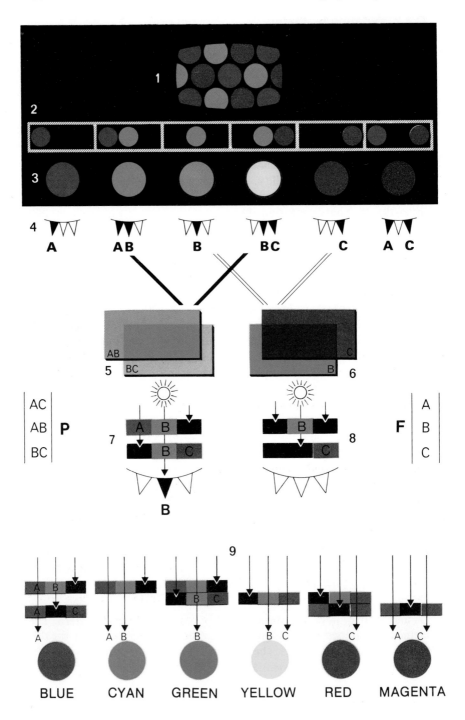

*Differential variables* 217

## C.4.3. Variation by color

Variation by color is very selective, but it cannot ensure complete selectivity any more than the other third-dimension visual variables can.

*What is color?*
Maxwell's theory gives a simple explanation for most of the observations we shall make here. It is as if the retina were lined with three types of cell: the A, B and C cones. When all three types are stimulated equally, we see "white." But when type A, for example, is less stimulated than the others, we see "yellow." Thus *color is an unequal stimulation of the cones.*

A color television screen is comprised of three kinds of points: blue, green, and red (1), even when the film is in black and white. The white image is thus due to the uniform stimulation of the three cones A, B, C.
- When the screen is blue, the green and red points are nearly invisible (2). Cones B and C are not stimulated (4).
- When the screen is yellow, the blue points are almost invisible, while the green and red points fill the entire screen. Cone A is not stimulated. In examining possible combinations, six cases are constructed (2), corresponding to the six colors in (3). In all six cases at least one of the three colors of the television screen is missing. The perceived colors are said to be "saturated" (4).
*A color is saturated when at least one of the three cones is not stimulated.* Conversely, "desaturation" simply involves stimulating, even weakly, the 3rd cone.

*"Binary" colors, and "primary" colors.*
A color is thus the blocking of one or two types of cone. It is a screen. A screen for two cones is a "binary" color* (**F**): blue, green, red. A screen for one cone is a "primary" color** (**P**): cyan, yellow, magenta.
The difference is simple . . . and essential.
*To superimpose two colors* is to superimpose two screens. If each only blocks a single cone, the third is not blocked, and the eye sees a color (5) and (7). On the other hand, if each blocks two cones, all three cones are blocked, and the eye sees "black" (6) and (8). In fact, the blocking or "occlusion" is never total, and the eye sees a brownish-gray.

*Sometimes called an "additive primary".
**Sometimes called a "subtractive primary".

**218**                                                          *The graphic sign system*

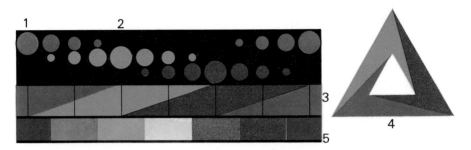
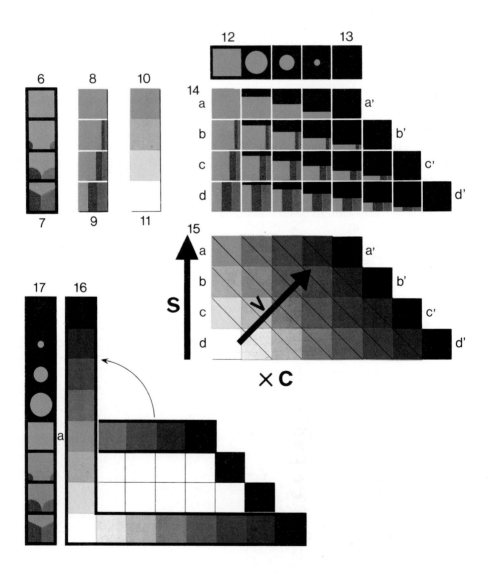

*Differential variables*

Consequently:
— the three "primaries," those which block only one cone, can be superimposed and can theoretically recreate all possible combinations of color **(9)**, p. 216.
Conversely:
— the three "binaries," those which block two cones, cannot be superimposed. But if they are *juxtaposed in such a way that the eye cannot separate them* (with a Newtonian disc or a television screen, for example), the eye can add them together and see the resulting colors.

*Color and saturation*

*The number of saturated colors is infinite.* In a "perfect" television screen, the passage from saturated blue **(1)** to saturated green **(2)** is a continuum. All the intermediates are saturated since the third cone, sensitive to red, is blocked. This is what is represented by schema **(3)**, whose cyclic form can be represented by Maxwell's triangle **(4)**.

*Desaturation of color.* Take a saturated green **(6)**. We can make it become more and more "white," and end up with a white with no coloration. To add white is in fact to add one of or both of the missing stimulations. It means going from **(6)** to **(7)**. This is *desaturation* of color, schematized by **(8)** → **(9)**. On a television screen, this color desaturation is independent of value. A "green" or a "blue" can be as bright as a "white."
But the same is not true on white paper. Here we must speak of value, that is distance in relation to white.

*Saturation and value.* On white paper, the desaturation of green corresponds to the gradation in **(10)** → **(11)**. Green becomes lighter and lighter. The same green **(12)** can also become darker and darker: **(12)** → **(13)**, ending up as "black." This progression is schematized by **(14a** → **a')** and corresponds to the gradation in **(15a** → **a')**. The same is true for each of the levels of desaturation b, c of green and also of white d, which can progress up to black b', c', d'. The set of gradations within green is schematized by **(14)** and **(15)**.

These figures enable us to distinguish the three dimensions of color.
1- The number of *hues* is infinite **(3)**.
2- The *saturation* of a given hue progresses from d to a.
3- On white paper, and for a given hue, the *value* progresses from white to black, according to the arrow V **(15)**.

The graphic sign system

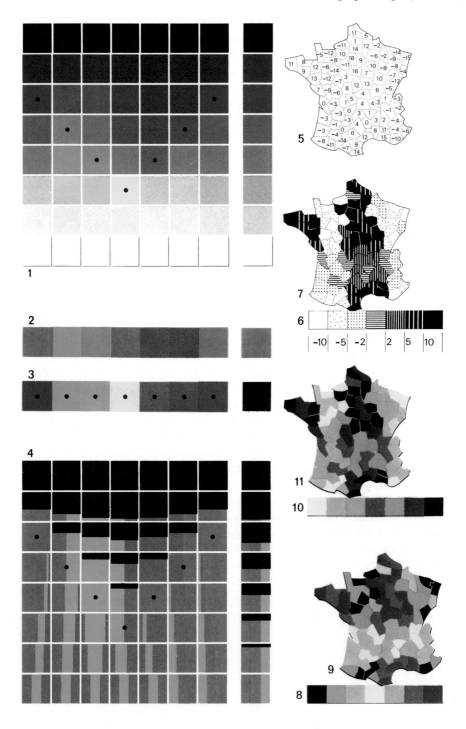

*Differential variables*

*Reduction to a two dimensional system.* In graphics, color is used to distinguish images. Therefore we stay as far away as possible from grays, and in general we exclude all the tones whose coloration disappears into grayness. In fact, we exclude the central part of figure **(15)**, p. 218, and we retain only the progression **(16)**, p. 218, for each color. This progression goes from white to black, passing through the saturated color (a), which on white paper corresponds to the "pure" color, the one neither "sullied" by black (darkened) nor "washed out" by white (desaturated).

*Color and Value*

The exclusion of gray shades enables us to construct table **(1)**\* which underlines three essential points in color graphics.

- *The pure colors afford optimum selectivity.* In fact, the farther we get from these colors, marked by black dots, the more the tones tend towards grayness.

- *The pure colors do not all have the same value.* In fact, the pure colors trace a V in table **(1)**. Aside from yellow, there are always two colors which have the same value, and the eye sees them as *similar* before seeing them as different.

Let us consider, for example, the information in **(5)**. Transcribed by values in **(6)**, it produces image **(7)**, a north-south image. Transcribed by the order of the spectrum **(8)**, it produces **(9)**, an east-west image. We note that it is impossible to disregard this orientation. It is as if the eye assimilates the two extremities of the spectrum as one unified perception, contrasted with the unity formed by the central colors. This is because the extremities are dark, whereas the central colors are light **(8** and **3)**.

- *Perception of values dominates perception of colors;* for example, a blue and a red of the same value are first seen as similar before being seen as different. If this proposition is true, the legend that orders the colors according to their value **(10)** will produce an image conforming to the distribution of the information. This is confirmed by image **(11)** and further confirms the observation made on p. 186.

Consequently:

1 - *Color selectivity varies with value.* For light values, optimum selectivity is obtained by green, yellow, and orange. For dark values, by red, blue, and violet **(1)**.

2 - *In the order of the spectrum, the pure colors create two visual groups,* and the representation of ordered data by spectrally ordered colors

---

\*Table **(4)** shows the theoretical proportion of each of the three cones A, B, C, for the tones in table **(1)**.

disturbs the display of information as we saw in **(9)**. In contour (isarithmic) maps, this disturbance can be negligible, since the two extremities of the spectrum are never brought together.

3 - *Colors of equal value are not visually ordered* **(2)**, and thus cannot represent ordered information.

*Defects of color*

Color differentiates images, thereby causing the neophyte to superimpose the very images that the use of color separated. However, as the complexity of a distribution increases, with or without color, regionalization becomes impossible and we are obliged to return to the elementary level of reading. Moreover, recent experiments in industrial surroundings show that a variation in sign orientation is generally more efficient than the use of color.

The author has the reputation of being against color. I am indeed against color when it masks incompetence; when it allows the superimposition of characteristics to the point of absurdity; when people believe it capable of representing ordered data; for as long as it wastes public funds by confusing knowledge with publicity; and for as long as it is manipulated by the media to convey false and illegible images.

Is color useful in pedagogy? Not always. In cartography, as well as in other fields, the modern school book would have us believe that using color is the only way to draw. Consequently, as an adult the designer believes only in color. He concentrates his efforts on the superimposition of numerous characteristics and loses sight of any possibility for regionalization, permutations and overall comparisons involving the entire set. He is merely doing encyclopedic work, and he demands color publication! It becomes a scientific and financial aberration when his demands are met. The drawing **(14)** on page 79 was published in color! Color is only indispensable in trichromatic analysis.

*Usefulness of color*

The three "primary" colors—"cyan-blue," yellow and "magenta-red"—can be combined to reconstitute all the colors. The superimposition of three characteristics, each in a primary color, enables us to see the regionalization produced by their overall structure. This is *trichromatic analysis*. It enables us to make full use of "map collections" (p. 162).

*Documents preparatory to research*, such as precision inventories and drawings on screens, make efficacious use of color. But beware! In these manuscript documents, blue and green or red and orange are easily confused. We must add a *difference in orientation* to the color differences

*Differential variables*                                           **223**

and maintain it throughout the entire figure. Only orientation admits no ambiguity, and thus avoids costly errors. Defining the color by the pencil number used is the quickest way to create total confusion!

*Publications involving topographical inventories:*—topographical, geological, or vegetation maps, ordered networks, anatomical drawings —depend on the complexity of the distribution. The procedure is simpler when the drawing juxtaposes different areas, as in geological maps. But whenever there is a superimposition within the same area, we must rely upon a difference in implantation—by point, or line, or area—since superimposed colors produce new colors.

To distinguish characteristics always remains difficult, often leading us to use a collection of maps, one per characteristic, published in reduced size and in black and white, in the margins of the colored map.

In the analysis of very similar values which nonetheless correspond to different characteristics (aerial photographs, thermographies), color provides the solution. We differentiate the degrees of value by appropriate technical means. Each degree is then colored so that it contrasts sharply with neighboring ones. Color thus enables us to differentiate meaningful values that the eye would only be able to distinguish with difficulty otherwise.

### C.4.4. Variation by orientation

A variation in orientation is especially useful in two cases:

- *In implantation by point it affords a selectivity often comparable to that of color* and always superior to that of shape. It is much easier to define a given group in (4, p. 171) than in (1, p. 170). Furthermore, oriented marks present no problem in drawing; it is enough to ensure strict parallelism within each orientation. Series (1) is the best. Avoid a 45° angle. Remember that inclined marks form a set in relation to orthogonal marks (7, p. 122). A variation in orientation combines very well with size variation.

- *In implantation by area* orientation is only selective with very coarse textures. *However, this differentiation is without ambiguity.* In a manuscript prepared with colored pencils, it is indispensable that color and orientation be associated, since manuscript colors are often ambiguous.

- *In implantation by line* we can only count on two selective orientations (2).

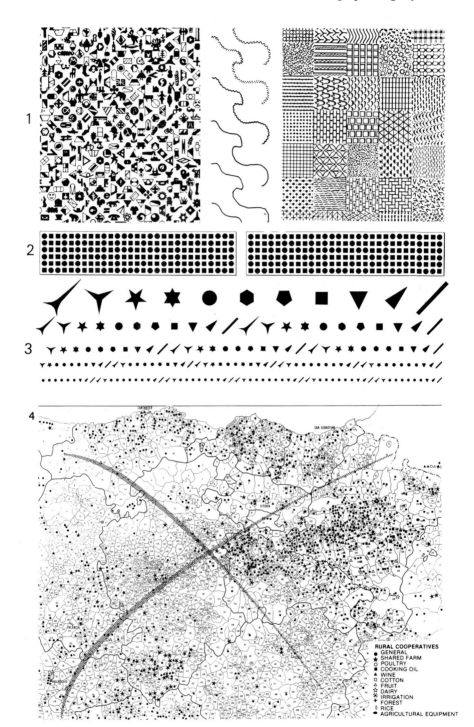

*Differential variables* 225

## C.4.5. Variation by shape

For all three implantations **(1)**, shape differences appear easy to draw and unlimited in number. Thus shape would seem to be the ideal variable for differentiating numerous characteristics. However, this is only true for answering elementary questions—"what is there at a given place?"—as with industrial drawings, working plans, and certain "mark maps."

But whenever a question of regionalization is pertinent, shape is no longer usable, since it has no selectivity whatsoever. It does not allow us, for example, to see the two words in **(2)**, that is, to distinguish the image formed by the circles. In intermediate or overall reading of the entire set, a square and a circle, all other things being equal, are practically similar.

Consequently, the use of differences in shape is strictly bound by very restrictive conditions. Furthermore, even an elementary shape is only identifiable above a certain size **(3)**. Finally, contrary to appearances, it is not easy to draw numerous differentiated shapes nor to copy any given shape rigorously, except the circle.

Shape variation is, in the end, very difficult to use correctly, and when the representational problems are inadequately analyzed, shape can become a major source of errors **(4)** that invalidate the graphic.

## C.4.6. Shape and color symbolism

How do we represent a factory? There is an infinite number of good solutions. As we saw on page 177, *this is not a problem for graphics*. Graphics is only concerned with the relationship *between* two things. External identification—"What is involved?"—is the province of verbal language and pictography. It establishes a more or less analogical convention between a sign and a thing. Is this convention, often necessary in mark cartography, subject to any laws?

*What is the aim of the conventional sign?*
- To stimulate the recognition that a factory is involved. This is an art!
- To save space; for example, by replacing the word "geodesic point" with a triangle.
- To save time, the time spent going back and forth to the legend or a dictionary, *provided the sign-word correspondence is memorizable*.

On the other hand, if memorization is impossible (with a legend of fifty different industries for example) the saving of space accomplished by the designer costs so much time for the reader that he turns away from the map. A map must save time.


The "standardization" of conventional signs therefore obeys:
1st - the fundamental law of graphics: not to destroy the relationships among the elements represented;
2nd - the laws of memorization:
- memorization is proportional to the repetition of the convention;
- it is inversely proportional to the number of conventions.


Consequently:
- Any attempt at standardization must conform to the correct representation of the relationships of resemblance, order or proportionality, *and to the absolute condition of not destroying them.* To standardize the square as a strong value or the circle as a weak value amounts to standardizing a lie. It is like proposing that $5 = 2$.
- As a function of repetition, any attempt at standardization can only be useful within a *restricted conceptual field capable of ensuring repetition* of the perception of the convention. When repetition reaches a certain threshold, the convention becomes a "symbol."
- The number of concepts being infinite, attempts at standardization can only involve a visual variable whose separable elements are of an infinite number. This variable is shape, but:
a) *Shape has no selectivity.* It excludes any possibility of regionalization. Therefore, it can only be useful for maps "read" point by point, that is "mark maps."
b) The infinite number of shapes favors the multiplication of conventions . . . *at the expense of memorizability.* The point of non-memorizability is reached very quickly, and this is "non-communication," that is, a useless drawing.
- Given the very restricted number of colors which are distinguishable by name (about eight), attempts at standardization of colors can only be applied within extremely limited fields.

*A few precautions to take in the use of conventional signs*

Any attempt at setting up conventions must avoid working against universal or quasi-universal constants.

*Physical constants linked to gravity:* The surface of water is not vertical; a walking man is not horizontal. A circle is less "stable" than a square. An arrow pointing down is derogatory.
Vertical and horizontal valuations are different. A vertical profile is better remembered than a horizontal one. Texture is more visible in a vertical line pattern. Oblique marks construct a set in relation to vertical and horizontal marks...

*Physical constants linked to morphology:* The correspondence between undulations and movement, between linear structure and stability. The unity of a curve, the multiplicity of a broken line. Discontinuity and punctuality of sand and vegetation. Continuity and linearity of water and clay... Parallelism and linearity of human constructions (roads); non-parallelism and sinuosity of natural elements (rivers)...

*Physical constants linked to natural colors:* rocks and bare ground are rarely green. Water, ice and vegetation rarely red. Heat and fire rarely blue.

*Physiological constants:* The differentiation thresholds of the steps of visual variables are inversely proportional to the size and value of the marks. Two juxtaposed marks are more readable than a single mark combining them, but five juxtaposed marks are unreadable. Red is the most pregnant of colors, that is the one which attracts the most attention, but *black is more visible than red*. "The heat" of red and "the coldness" of blue are real, but blue can be a particularly hot color. Blue "goes backwards," red "goes forwards," but the opposite may also be observed. Value relationships among colored areas vary with the light source.

*Quasi-universal socio-cultural habits:* We do not represent a reduction or a negation by +, an increase or an affirmation by −. Time moves from left to right . . . etc.

Let us note that most of these precautions are negative. With symbolism "it is better not to do such and such." That is about all we can say. Symbolism is an irrational language (any proposition can be questioned), superimposed on the logical language of graphics. Consequently, symbolism can only be efficacious and capable of replacing words if it becomes a habit, sustained by the constant repetition of a restricted codification and applied to a very limited field. The real problem in the standardization of conventional signs is not to define shapes but *to define limited fields of application*.

The greatest difficulty thus stems from the contradiction between standardization, a restrictive operation, and modern scientific impulse, which tends to break up fields in order to discover more significant and useful sets.

## C.5. THE LAW OF VISIBILITY

A construction can be correct and yet produce an invisible image of its elements are not differentiated among themselves. In a matrix, a completely black row does not separate the objects; it is not "differential." We eliminate it because it is useless and harmful to perception. This underscores the importance of the "law of visibility," a major condition of effective graphics:
*Any non-differential element is useless and reduces the visibility of the image.*
Consequently we must:
a) *eliminate what is common* to all elements of the information; common features have no differential qualities;
b) *use the entire length of the visual variable* in order to transcribe the information; for example, use a range of values from white to black. In a range from gray to black, the white-gray variation is not differential. It is useless and reduces the visual distance between the differential elements.
c) *eliminate the "noises,"* that is the non-signifying perceptions: gray or colored paper, dirty paper, spots, jumbles, graphic confusion, etc. They also reduce visual differentiation.

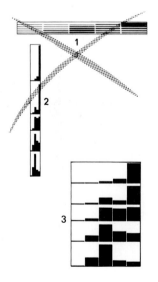
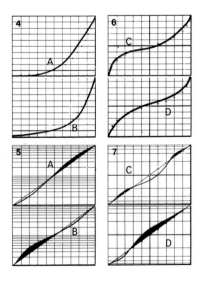

# The law of visibility

*Several applications*
In **(1)** the *y* variable of the plane is not used over its entire length in relation to *x*. In **(2)** the opposite is true. In both cases the meaningful form is invisible. We must aim towards a square plane **(3)**.
In **(4)** the two curves have one part in common: a tendency towards a geometric progression. The logarithmic scale **(5)** neutralizes this tendency by linearizing it. The differences between the curves become visible and are highlighted by shading the positive distances from the straight line **(5)**.
The same is true when we have to compare "repartitions" which are close to Gauss's "normal" law **(6)**. But here it is the "Gaussian" scale **(7)** which neutralizes the common tendency. In a series of ordinary curves, the mean curve enables us to neutralize the common tendencies.

The meaningful form is much more visible in **(9)** and **(10)** than in **(8)**. **(9)** and **(10)** eliminate the useless common parts. Here are three ways of achieving this:
a) Shade as in **(9)** the numbers which exceed the mean of the line.
b) Shade as in **(9)** starting from the largest numbers. Stop when the neighboring profiles resemble each other. The numbers thus discovered characterize the correlations among the profiles.
c) In each row let white correspond to the lowest number and black to the highest number **(10)**. Divide arithmetically. We can combine *c* with *a* or *b*. Note that **(10)** shows more detail than **(9)**. The image-file **(12)** follows the principle in *c*. It is much more differential than **(11)**. But for a very large image-file, the best solution is **(13)**, which only shades that which exceeds the mean line.

Diagrams of type **(14)** can be useful in certain mark maps. We must then draw **(15)**, which becomes differential by highlighting that which exceeds the general mean of each orientation.

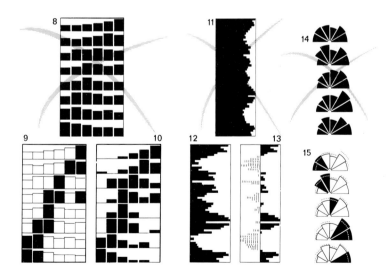

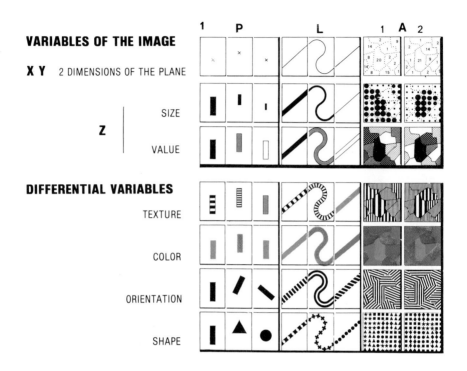

## C.6. SUMMARY

A. To understand is to reduce the multitude of elementary data by discovering similar elements, higher sets and relationships. This is the main problem in information processing. Any problem, conceivable through a "matrix analysis," is or can be written in the form of an $x\,y\,z$ data table. To understand is to discover the $x$ and $y$ groupings constructed by the $z$ data.

B. To transcribe this $x\,y\,z$ table, graphics possesses eight visual variables **(1)**. But to solve problems of information processing, the limits of the image lead to the following constructions.

C. *Diagrams* transcribe the relationships between two sets of elements. To solve our problem, that is to discover the groupings along $x$ and $y$, it is obviously necessary to "copy" the data table. This means using the $x\,y$ *matrix construction* **(4)** and obtaining *groupings through permutations*.

# Summary

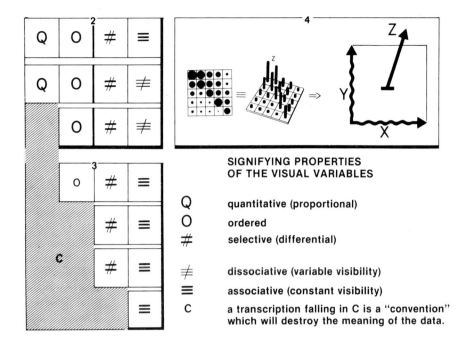

**SIGNIFYING PROPERTIES OF THE VISUAL VARIABLES**

- Q    quantitative (proportional)
- O    ordered
- #    selective (differential)
- ≠    dissociative (variable visibility)
- ≡    associative (constant visibility)
- C    a transcription falling in C is a "convention" which will destroy the meaning of the data.

Apart from the exceptions discussed in E and F, no other construction will solve the problem.

D. When the $z$ data from the table are ordered or quantitative, they are transcribed by value or size, the only ordered variables in $z$ **(2)**. They create the meaningful form. The other third-dimension variables **(3)** are not ordered*, and will not construct a meaningful overall image of the entire set. They will not solve our problem.

E. When a quantitative table has *fewer than four rows, a scatter plot* transcribes one row along $x$, one along $y$, the third in $z$ and displays the sought-after groupings, without permutations.

F. In a table comprising numerous yes/no characteristics, it may be preferable to construct a *collection of reclassable tables*. The point of comparison is provided by the correlation (in a scatter plot) of two ordered characteristics.

---

*Except for texture, but its constant visibility constructs a poorly differentiated image.

G. *Networks* transcribe the relationships among the elements of a single set.

H. *Reorderable networks* (e.g. flow charts) utilize the plane to transcribe elements and their relationships. Networks are difficult to transform and only make groupings appear with a limited number of elements. However, by transcribing the elements once along $x$ and once along $y$, we return to a permutable matrix.

J. *Ordered networks or "topographies"* (e.g. maps) copy the arrangement of elements in a physical object: a machine, a living being, geography . . . .They derive their properties from the constancy of this arrangement.

K. Transcribing an ordered or quantitative characteristic as a network conforms to law D.

L. Transcribing several characteristics as a network, on a map, for example, raises the problem of the selection of characteristics. It has two imperfect solutions:
*a)* when we are to define groups, the pertinent question is "a given characteristic, where is it?" To reply we must construct one map per characteristic. The collection of maps is reclassable, but *b)* has no visual answer.
*b)* when the map serves for spatial marking, the pertinent question is "at a given place, what is there?" We must *superimpose the characteristics* on a map and utilize the differential visual variables **(3)**, but *a)* has no visual answer.
To answer all questions, we must construct *a)* and *b)*.

M. *In a superimposition* of characteristics, selectivity increases
*a)* when we contrast the three implantations: by points (P), lines (L) and areas (A);
*b)* when we *combine several variables*. Note then that any combination possesses the properties of the highest variable in the table **(2, 3** on p. 231), except for the size/value combination, which can be concurrent or compensating (*S.G.*, p. 186).

# D. THE MATRIX ANALYSIS OF A PROBLEM AND THE CONCEPTION OF A DATA TABLE*

Traditional mathematics teaches us to solve equations.
Modern mathematics teaches us to set up these equations.
Can modern graphics also develop in this direction?

Once the data table is constructed, we now know how to process it, mathematically or graphically, with reference to our "synoptic."
But what sort of data table should be constructed? This is the new and basic problem. And experience and logic convince us that this problem is also of a matrical nature. A study is a whole entity that constitutes one and only one matrix. Clearly, for relationships among a study's different chapters to be logically demonstrable, a single point in common does not suffice. *A point of comparison is also necessary;* otherwise, the relationship among these chapters would be merely verbal and would not provide the elements necessary for a logical demonstration.
Matrix analysis is a process of reflection based on graphic notation. It enables us, whatever the magnitude of the problem, to conceive it as a single table, starting from which all necessary choices can be logically discussed and connected together. This connection establishes the real chapters of the study.
This is a recent procedure, constantly offering new and fascinating developments, and can only be presented here in its simplest form.

*See p. 17.

The conception of a data table which will be transcribed graphically or supplied to the computer depends on:
- the available or desirable information;
- the hypotheses, that is the pertinent questions;
- the available time and the accessible processing means.

The matrix analysis of a study visualizes the entire set of information in such a way that the investigator *can define a data table* compatible with a processing method capable of producing an answer to the pertinent questions and also compatible with the available means.

Matrix analysis enables us:
- to define the point of comparison and *to ensure the homogeneity* of the study;
- to "dimension" the problem in relation to the means;
- *to record the questions* in a manner appropriate to the data and to ensure that the processing method will answer these questions;
- to alleviate the ambiguity of the statistical presentation and *to display the information* being considered in a precise and concise way for potential users: advisers, mathematicians, designers, computer programmers;
- to display the information considered initially and thus *justify the choices and processing* methods retained in the subsequent study.

In order to respond to a wide variety of statistical situations, matrix analysis comprises the successive drafting of three documents:
- *The apportionment table* makes a precise inventory of all the components taken into consideration and all their intersections. It avoids the ambiguity of a statistical presentation.
- *The homogeneity schema* outlines the table from the information. It enables us to define a homogeneous table, to raise problems of processing methodology, and to conceive a utilizable data table.
- *The pertinency table* enables us to record the hypotheses opposite the final table, and to conceive new data or "derived" data (sums, ratios) more pertinent to the hypotheses, and finally to record all the calculations used in preparing the final data table.

*The apportionment table*

## D.1. THE APPORTIONMENT TABLE

This table has three properties; it must:
- record all the components of the study;
- record the length of each component and its intersections with other components;
- be independent of the order of the text.

### D.1.1. Inventory of the components

The table is set up on a large sheet of paper ruled in squares. In a large column on the left we record successively by order of appearance all the components of the text.

This is not as simple as it would appear. In fact, experience shows that the correct identification of the components can pose a serious problem for the researcher. If he does not know how to display his data clearly, how can he process them appropriately? Take, for example, a study of professional florists. What can we make of the following statement: "I know the nurserymen, the merchants, the wholesalers of cut flowers, the producers specializing in bulbs."
Such a statement immediately demonstrates the weakness of the logical analysis. What are the *differential concepts* and what sets do they share? Is it true that it concerns:
a) a set of professionals;
b) divided into producers, merchants, miscellaneous;
c) divided into mass producers and specialists;
d) the subset merchants divided into wholesalers and retailers;
e) merchandise divided into nursery, cut flowers, bulbs, miscellaneous?
The apportionment table forces us to define the components **(1)**. When the study is complex, the table is not easy to set up. But it is precisely then that it becomes indispensable.

- *Professionals*
- *Types of profession: producers, merchants, miscellaneous*
- *Mass production: yes, no*
- *Types of merchant: wholesaler, retailer*
- *Types of product: nursery, cut flowers, bulbs, miscellaneous.*

# Matrix analysis

Let us return to a simpler example, as in the statement: "I know the population by profession and by age." This statement can correspond to completely different types of information. Simple questions enable us to clarify the situation.

### D.1.2. Recording intersections

Do we know the profession *of each individual?* Yes! How many individuals, how many professions? The answers are recorded as in (1).
*Two crosses in the same column define two components which intersect and construct a double-entry table,* as schematized in (2). Recording the lengths of the components enables us to define the dimensions of the table: 250 × 10 = 2500 cells. 01 signifies yes/no answers.

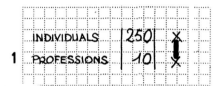 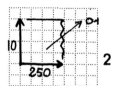

Do we know the age of *each individual*? No! We know his age group. How many groups? Four! This answer is added to the preceding one and is recorded as in (3).
*Two crosses in the same row define a component common to the two corresponding tables* (4). Dimensions: 250 × (10 + 4) = 3500 cells.

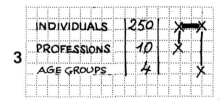 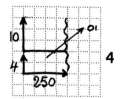

When the age of *each individual* is known, it is expressed as a quantity of years. One row is needed to write the numbers, and we record this as in (5).
*The notation Q defines a quantitative component whose quantities are*

*expressed in z. When there is only one cross in the column, the table has only one row* **(6)**. Dimensions: 250 × 1 = 250 cells.

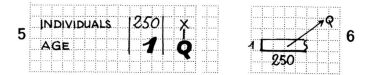

Do we know the age and profession *of each individual*? No, but we know the number of individuals per profession for each age group. This quantity is written in *z* as in **(7)**.
*The notation Q defines a quantitative component expressed in z for a table defined by the crosses appearing in the column* **(8)**. Dimensions: 10 × 4 = 40 cells. Qob signifies quantity of "statistical objects."

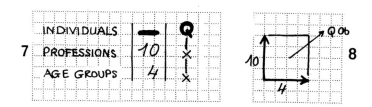

We know the profession *of each individual*. But for age we only know the number of individuals per age group. The "individuals" component provides in the first case, an entry in the table and in the second a quantity expressed in *z*, as recorded in **(9)**.
*The presence of X and Q in the same row defines a component which provides:*
- *for X, an entry in the table* **(10)**;
- *for Q, quantities of objects, expressed in z in another table* **(11)**.
Dimensions: 10 × (250 + 4) = 2540 cells.

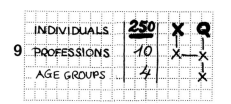 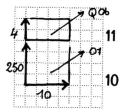

*Matrix analysis*

## The apportionment table

All these notations, as well as subsequent ones, are independent of the order of the signs, either in row or column. They are also independent of the order of the text.

In order to define all possible statistical situations, we must add two signs.

An ordered component, age groups, for example, can be transcribed by the ordinal number of each group (2). This only activates a single row of the table. In this case Q is replaced by O, and recorded as in (1).
*The notation O designates the presence of ordinal numbers.*

The notation ╲ designates the presence of empty cells or of data common to several elements.

*Data in common.* Take the components in (3). We know the profession of each individual. We also know the material characterizing the building each one inhabits. The material thus characterizes at once both the building and the individuals inhabiting it.

*Data inapplicable to certain categories of a component.* Take the components in (5). The professions obviously do not apply to children. We use a special notation, as in (5), to designate an "inapplicable" area (N.L.) in the table (6). In this case the apportionment table (5) must list all the categories or groups of the component, by reserving a row for each of them. The specific categories involved can thus be clearly identified.

*Missing data.* An apportionment table (7) shows that the age of 250 individuals is known but that we only know the profession for 100 of them. We must reserve a row for noting the intersections concerning this subgroup of individuals. Information of this kind creates an unknown area in the homogeneity schema. We will see later (p. 256) that this area can be of fundamental importance. It is the basis for all processing by "interpolation."

## D.2. THE HOMOGENEITY SCHEMA

The apportionment tables **(1)** to **(7)** define the homogeneity schemas **(8)** to **(14)**. I: individuals, P: professions, R: income, A: age.

### The z notations

- 01 (or z01) defines a table whose cells contain only yes/no answers.
- Q (or zQ) defines a table containing only various quantities and/or orders.
- Qob (or zQob) defines a table containing only quantities of a single population of statistical objects, that is, whose general total is meaningful.
- But one table can contain all or some of these statistical situations. We then note zD or z "diverse" (p. 254).

**D.2.1.** *The homogeneity schema is the* x y z *schema of the double-entry table constructed by the data.* It is independent of the number of components: we can as easily record 250 as 10 million.
The apportionment table enables us to record the content of the

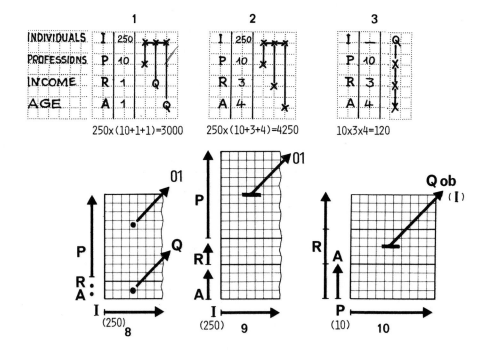

## The homogeneity schema

statistical text linearly and in any order. When the text is complete—and this should be stated—the apportionment table enables us to discover and schematize the largest table which can result from the entire data set.

*The relationship between the apportionment table and the homogeneity schema*

What is the aim of this analysis? To define a point of comparison for the entire study, that is to find the component along $x$ which enables all the other components in the study to be placed along $y$. In other words, to discover the set of "statistical objects" of which all the other components are attributes. This component is immediately obtained from the apportionment table. It is the horizontal row of crosses. When there are several of them we must study various schemas, paying particular attention to empty areas constructed by each of them (pp. 239 and 256). When there is only one vertical line **(3)**, each of the components with a cross can be placed along $x$ **(10)**. And when there are several vertical lines but no horizontal row of crosses **(7)**, we have a heterogeneous set, that is one without a point of comparison **(14)**.

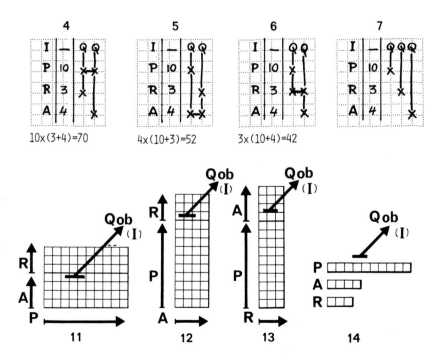

# Matrix analysis

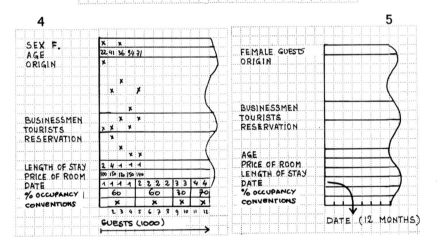

*The honogeneity schema* 243

## D.2.2. The homogeneity schema for the hotel example

What are the implicit or explicit operations that enabled us to set up the data table shown at the beginning of this book?
From the statistics contained in the hotel records **(1)**, the assistant compiled a statistical card **(2)**, that is a card comprised only of cells with yes/no answers, quantities, or ordinal numbers (dates). He estimated the total number of guests to be about 1000, represented by 1000 statistical cards **(2)**. Next he made a complete inventory of useful data. This is the *apportionment table* **(3)**. From it he derives the *homogeneity schema* **(4)**, that is the type of table capable of containing all the data prior to any calculation of aggregation.
This table has $18 \times 1000 = 18000$ cells and must be reduced.

*The homogeneity schema* reveals the necessity of processing the data and organizes the possible choices.
We know (pp. 18 and 251) that there are four solutions:
1) Direct graphic processing, impossible here;
2) Retaining the component "guests" and using samples;
3) Retaining the component "guests" and using multivariate analyses;
4) Choosing a new point of comparison and aggregating on this basis, that is adding the objects, here the "guests."
In this last case *any component placed along y in the schema can serve to regroup the objects placed along* x. The homogeneity schema thus displays all the possible choices.

What are the characteristics of the slow periods? This question, raised from the start, obviously leads to choosing the component "dates" to aggregate table **(4)**. But into 365 days, fifty-two weeks, twelve months or four quarters? We decide on months **(5)**. It now remains to determine which calculations will enable us to construct the final table and to verify that all the indicators retained are comparable and meaningful. This is the aim of the *pertinency table*.

244                                                                                 *Matrix analysis*

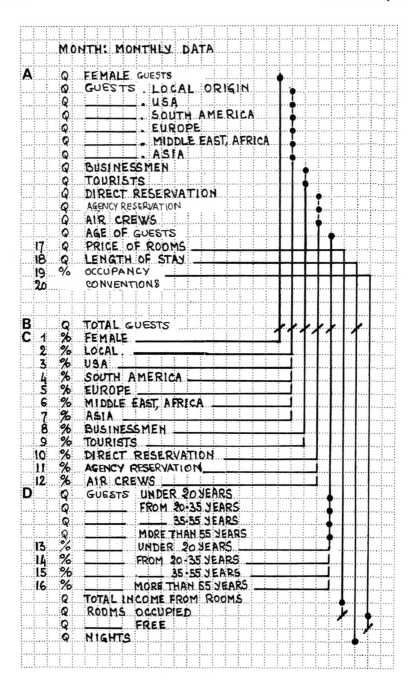

# D.3. THE PERTINENCY TABLE

## D.3.1. The hotel example

To organize the calculations in the hotel example, we compiled the data list **(A)** as it was defined by the last homogeneity schema (p. 242).

- Should we retain the total quantities or calculate percentages? Percentages eliminate a constant and are thus preferable. We must therefore calculate the total number of guests per month and divide each category of guest by this total. This is notation **(B)**, from which we derive the indicators numbered from 1-12 **(C)**. It is convenient to note (●) number of clients, (/) divided by the total number. We can also use (●) quantity A, (−) less quantity B, or (+) plus quantity C.
- Should we calculate an average monthly age or use age classes? Classes are much more meaningful. Their choice **(D)** leads to calculations that we record and that result in indicators 13-16.
- To calculate the average monthly price of a room **(17)**, we divide the total income from rooms by the number of rooms occupied.
- For the average length of stay **(18)**, we calculate the total number of "nights" and divide it by the total number of guests.
- For the percentage of occupancy **(19)**, we divide the total number of rooms available by the total number of rooms occupied.

The indicators which will be recorded on the data table **(E)** are numbered. We see that for the most part they are derived indicators. The pertinency table enables us:
- to define the pertinent indicators precisely;
- to define the necessary calculations and organize them in an economical way. But it also enables us to record hypotheses and even to generate them.

**E**

| J | F | M | A | M | J | J | A | S | O | N | D | | |
|---|---|---|---|---|---|---|---|---|---|---|---|---|---|
| 26 | 21 | 26 | 28 | 20 | 20 | 20 | 20 | 40 | 15 | 40 | | 1 | % GUESTS FEMALE |
| 69 | 70 | 77 | 71 | 37 | 36 | 39 | 39 | 55 | 60 | 68 | 72 | 2 | % —//— LOCAL |
| 7 | 6 | 3 | 6 | 23 | 14 | 19 | 14 | 9 | 6 | 8 | 8 | 3 | % —//— U.S.A. |
| 0 | 0 | 0 | 0 | 8 | 6 | 6 | 4 | 2 | 12 | 0 | 0 | 4 | % —//— SOUTH AMERICA |
| 20 | 15 | 14 | 15 | 23 | 27 | 22 | 30 | 27 | 19 | 19 | 17 | 5 | % —//— EUROPE |
| 1 | 0 | 0 | 8 | 6 | 4 | 6 | 4 | 2 | 1 | 0 | 1 | 6 | % —//— M.EAST, AFRICA |
| 3 | 10 | 6 | 0 | 3 | 13 | 8 | 9 | 5 | 2 | 5 | 2 | 7 | % —//— ASIA |
| 78 | 80 | 85 | 86 | 85 | 87 | 70 | 76 | 87 | 87 | 80 | | 8 | % BUSINESSMEN |
| 22 | 20 | 15 | 14 | 15 | 13 | 30 | 24 | 13 | 15 | 13 | 20 | 9 | % TOURISTS |
| 70 | 70 | 75 | 74 | 69 | 68 | 74 | 75 | 68 | 64 | 75 | | 10 | % DIRECT RESERVATIONS |
| 20 | 18 | 19 | 17 | 27 | 27 | 19 | 19 | 26 | 27 | 21 | 15 | 11 | % AGENCY —//— |
| 10 | 12 | 6 | 9 | 4 | 5 | 7 | 6 | 6 | 5 | 15 | 10 | 12 | % AIR CREWS |
| 2 | 2 | 4 | 2 | 2 | 1 | 1 | 2 | 2 | 4 | 2 | 5 | 13 | % GUESTS UNDER 20 YRS. |
| 25 | 27 | 37 | 35 | 25 | 25 | 27 | 28 | 24 | 30 | 24 | 30 | 14 | % —//— 20-35 —//— |
| 48 | 49 | 42 | 48 | 54 | 55 | 53 | 51 | 55 | 46 | 55 | 43 | 15 | % —//— 35-55 —//— |
| 25 | 22 | 17 | 15 | 19 | 19 | 19 | 19 | 20 | 19 | 22 | | 16 | % —//— MORE THAN 55 —//— |
| 163 | 167 | 166 | 174 | 152 | 155 | 145 | 170 | 157 | 174 | 165 | 156 | 17 | PRICE OF ROOMS |
| 1.65 | 1.71 | 1.65 | 1.91 | 1.90 | 2. | | 1.54 | 1.60 | 1.73 | 1.82 | 1.66 | 1.44 | 18 | LENGTH OF STAY |
| 67 | 82 | 70 | 93 | 74 | 77 | | 56 | 62 | 90 | 92 | 78 | 55 | 19 | % OCCUPANCY |
| | | | x | x | x | | | x | x | x | x | 20 | CONVENTIONS |

## D.3.2. Derived data

Consider, from a regional planning study, list **(1)**, which results from a preliminary homogeneity analysis.

In setting up the pertinency table **(2)**, it is useful to reclass the data. The principle is simple: to write the least number of words possible and replace repetitions by lines. Reading becomes much faster and facilitates further operations.

*Derived data.* The available data are in "absolute quantities." We know that they must be replaced by proportions; otherwise, the largest communities in matrices or maps will always be "black," the smallest "white." Table **(3)** indicates all the necessary calculations.

At this stage, we can imagine new indicators, and, if imagination falters, the pertinency table enables us to bolster it.

## D.3.3. Recording hypotheses

It is in fact not certain that all the data items will be useful. Neither is it certain that all the useful data have been derived. But in this form, the table enables us to record hypotheses graphically, that is to show the questions that initially defined the problem along with potential new questions. Here are some examples.

*First hypothesis (H1):* Is it true that population size is accompanied by an increase in telephone and retailer density and corresponds with population growth? In column H1 on the left of table **(4)**, all the elements of this question are recorded: + (the greater) the population, + (the greater) the percentage of retailers... and we discover that we must calculate population growth **(21)**.

*Second hypothesis (H2)*: Is is true that the percentage of the elderly increases in inverse proportion to the tertiary population and in direct proportion to grain crops? In column H2 we note: + (the greater) the population of elderly, − (the smaller) the tertiary population ... and we discover that we must calculate the tertiary population, which implies checking whether the total population is indeed the sum of agricultural, industrial and tertiary populations. If this is the case, it is sufficient to deduct the sum of the non-tertiary percentages from 100. This is noted in column (D1). The calculation of the area devoted to grain crops is carried out in two stages: the amount of area (D2) and then the percentage of area (D3).

# The pertinency table

**1**
- Q POPULATION 1968
- Q POPULATION 1954
- Q AREA OF THE COMMUNITY
- Q POPULATION AGRICULTURAL 1968
- Q POPULATION INDUSTRIAL 1968
- Q AREA AGRICULTURAL 1968
- Q AREA FOREST 1968
- Q AREA GRASS 1968
- Q TELEPHONES 1968
- MAIN ROAD
- Q RETAILERS
- Q POPULATION +60 YEARS OLD 1968

**2**

| | | | | |
|---|---|---|---|---|
| 1 | Q | AREA: | TOTAL | 1968 |
| 2 | Q | | FORESTS | |
| 3 | Q | | AGRICULTURAL | |
| 4 | Q | | GRASS | |
| 5 | Q | POPULATION | TOTAL | 1954 |
| 6 | Q | | | 1968 |
| 7 | Q | | AGRICULTURAL | |
| 8 | Q | | INDUSTRIAL | |
| 9 | Q | | +60 YEARS | |
| 10 | Q | RETAILERS | | |
| 11 | Q | TELEPHONES | | |
| 12 | MAIN ROAD | | | |

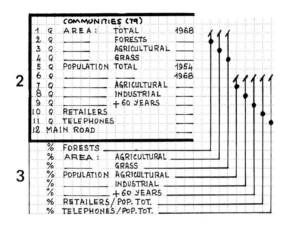

**3**

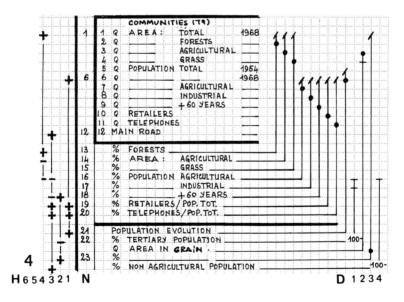

**4**

*Third hypothesis (H3):* Does the presence of a main road increase (+) the development of retail trade, telephone usage, non-agricultural population, and does it decrease (−) the number of the elderly? To answer these questions one might think that calculating the percentage of nonagricultural population is necessary. However, this is a useless calculation since this component is the complement of the agricultural population. At the moment of graphic processing two solutions are possible: to visually perceive that the agricultural population is classed in inverse order to the (+) components and in corresponding order to the (−) component (column H3); or just as well, to invert the "agricultural population" row (which is possible with certain permutation equipment whose rows can be turned over, reversing their order). This inverse order corresponds here to the "nonagricultural population."

*Fourth hypothesis (H4):* Is it true that the largest communities are forested and, conversely, that in these communities the agricultural area and population tend to be less? Here we have all the necessary indicators.

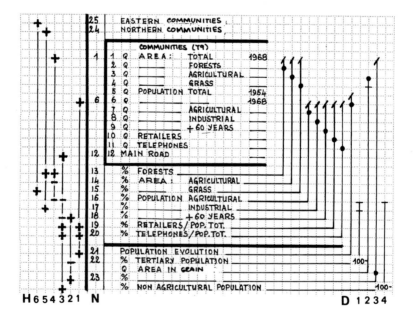

# The pertinency table

*Fifth hypothesis (H5):* Is it true that the forested communities are situated to the north of the zone being considered and that they favor an industrial population? This hypothesis leads us to define, according to a map, a new indicator, northern communities, and to record it **(24)**.

*Sixth hypothesis (H6):* The notation (H6) enables the reader to formulate this hypothesis, which introduces a new indicator: eastern communities.

### D.3.4. The final table. Two classic questions.

The final table retains only the data corresponding to the pertinent questions. These data are identified by numbering them in column (N), and generally they are calculated data.

*Which numbers should be recorded in the statistics?* We should only record those data which cannot be calculated, that is absolute quantities of the parts. The "part" sums, the differences, the percentages and the necessary ratios, defined by the pertinency table, are calculated afterwards by a calculating machine or computer.

*In what direction do we calculate percentages and construct profiles?* Percentages make the objects comparable by reducing the totals per object to 100. Consequently:

*when the objects are along* x **(2)**, *the totals* **(3)**, *the 100s* **(4)**, *and the profiles* **(5)** *are along* x.

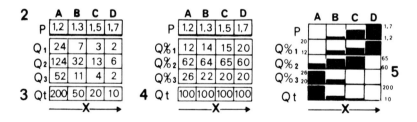

When the objects are in z, that is when the data tables are of the zQob form (p. 255), the direction of 100 does not matter (p. 62). This would be the case in **(2)** if the table was comprised only of rows Q1, Q2, and Q3, that is only the addable rows.

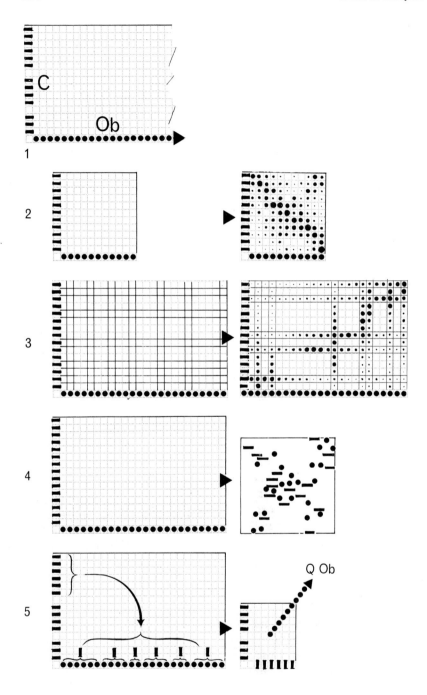

*Applications* **251**

*D.4. APPLICATIONS OF MATRIX ANALYSIS*

## D.4.1. Choice of procedure and point of comparison

Once the homogeneity schema is constructed, what are the available processing methods?

Consider a classic example: a survey yields a collection of questionnaires to be analyzed. The homogeneity schema is set up as in **(1)**, that is with the objects (individuals for example) along $x$ and the characteristics along $y$.* This schema describes the matrix of available information. The choice of method now depends on three factors: the pertinent questions (hypotheses), the dimensions of the matrix, and the time and means available. It is in relation to these three factors that we will choose from among four methods.

*Direct graphic processing* **(2)** is possible when the data matrix does not exceed $100 \times 100$. It enables us, as in the hotel example, to reduce $x$ and $y$ or, as in the "wood-lice" example from page 75, to study each of the components.
Graphic processing is also adaptable to other dimensions, depending on the nature of the data (see the synoptic, p. 31). If the dimensions of the matrix exclude direct graphic processing, there are still three possibilities.

*Sampling and interpolation* **(3)** only process part of the data. We retain all the characteristics for a reduced number of objects and all the objects for a reduced number of characteristics. The problem is to determine representative elements (p. 259).

*Multivariate analyses* **(4)**—factor analysis, hierarchical cluster analysis and discriminant analysis enable us to undertake investigations which would be unthinkable without a computer, namely to successively compare each object with all the others, each characteristic with all the others, and thus define "distances" and groups.
But here also choices are necessary: the choice of an algorithm, of a distance metric, of subsets to be processed separately for easier inter-

---

*This is contrary to conventional American usage. However, objects can be defined by numbers, whereas characteristics must be defined verbally. To read these definitions easily, it is advisable to write them horizontally and thus to record the characteristics along $y$.

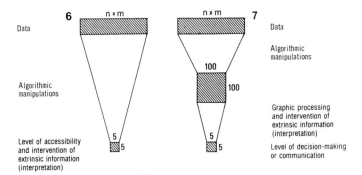

pretation. Indeed, interpreting the results also poses problems because the basic information, that is the data matrix, has disappeared, taking with it the elements needed for precise interpretation and the means of defining subsets that often lead to the most original observations. We note, moreover, that these calculations also have their own limits, certainly broader than those of graphics but nevertheless of the same order of magnitude in comparison with the scope of modern information resources.

We note in conclusion that all these algorithms end in graphically expressed results and that they prefigure a future solution: the possibility of using mathematical and graphical means of information-processing at will. This solution means replacing schema **(6)** by schema **(7)**, enabling us to introduce the essential element of interpretation, namely extrinsic information, at a less simplistic level than in present solutions (100 and five are merely orders of magnitude).

*Aggregations* **(5,** p. 250) that is *additions*. The $x$ dimension, equal to the number of objects or individuals, can be reduced by grouping individuals according to community or profession for example. *Any characteristic placed along* y *in the homogeneity schema can serve to group objects placed along* x. But be careful! In choosing a characteristic for grouping objects, *we construct a new point of comparison* (p. 243).
To aggregate the data is to transform the problem. It means answering

*Applications* 253

other questions, moving toward another system of comparison. In choosing the characteristic to put along *x*, we choose a discipline. The historian chooses time, the geographer regions, the sociologist social groups, the psychologist individuals . . . The image has only three dimensions, and this is no doubt why different disciplines exist. In any case this transformation is a given in the preceding processing methods, which have the property of preserving the initial point of comparison, and thus do not transform the problem.

Is it necessary to aggregate the data? Do the hypotheses enable us to change the point of comparison, and if so, which one should be chosen? If not, what processing method should be adopted? All the elements of this fundamental discussion emerge from the various tables and documents involved in matrix analysis.

### D.4.2. Brief description of the aggregation cycle

Consider again the data on page 240. They allow us to identify the three successive tables which define the aggregation cycle.

*The cycle*
Table (1) supplies the data: "objects," here individuals (I), for whom we know three characteristics: age (A), income (R) and profession (P).
Table (2) transforms ages and incomes into age classes and income classes, enabling us to construct a table with only binary entries, necessary for the following operation.
Table (3) portrays the objects in *z*. The quantities of individuals become the measure of the relationships of age, income and profession. The dimensions of the table are independent of the number of objects. They are simply a function of the number and length (number of classes or categories) of the characteristics. The component "objects" is the point

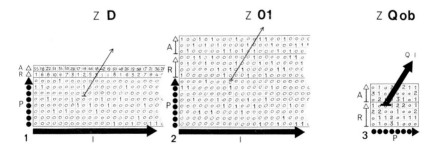

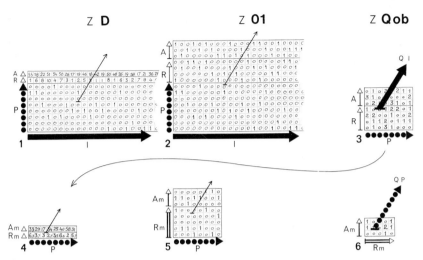

in common, but it is no longer the point of comparison, now supplied by the "professions."

From **(3)** we can begin a new cycle in which the "professions" (P) have become the "statistical objects." Figure **(4)** gives the mean age (Am) and mean income (Rm) per profession. **(5)** transforms these numbers into three age classes and five income classes. **(6)** portrays the professions in $z$. The quantities of "professions" measure the relationship between age and income . . .
And so on, as a function of the number and length of the components, until the information is exhausted.

*Three aggregation tables*

They follow each other in succession and can be defined by the nature of $z$: zD, that is "diverse" $z$, in which the cells can contain quantities, ordinal numbers or binary answers (01); z01 which contains only binary answers; zQob in which the cells enumerate the objects.

In this sequence of transformations the COMPONENT "professions" for example:
- is a CHARACTERISTIC attributed to individuals **(1)** and **(2)**;
- becomes a set of STATISTICAL OBJECTS to which characteristics **(3)** to **(5)** are attributed;
- is finally a QUANTITY OF OBJECTS measuring the relationships among characteristics **(6)**.

# Applications

The notions of statistical object, characteristic and quantity only take on meaning within a dynamic conception of the aggregation cycle. These are matrical notions depending only on the finite set of data being considered and on their level of aggregation.

*Primary application*
When the data come from different tables, each table is situated at a certain level in the aggregation cycle of the new homogeneous table containing all the data. Consequently, in order to reconstitute a homogeneous set providing a point of comparison, we must reconstitute the *schema* of the new table.
It then enables us to see if we can trace back to the original data, and if this is impossible, to determine *at what aggregation level and on which "statistical object"* the homogeneous table may be realizable.

*Four forms of statistical tables*
The aggregation cycle **(7)**\* starts with two tables of the "Characteristics/Objects" form, differentiated by $z$. It ends up in a "Characteristics 2, 3 . . ./Characteristic 1" table.
*Calculating the deviations (d)* constructs two other forms of table **(8)**, filled up only to the diagonal: *set of characteristics* on itself, *set of objects* on itself. This last table also denotes the relationships ($r$) among objects in a network **(20,** p. 128). All these tables are permutable.

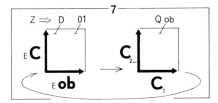 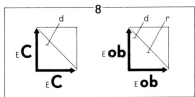

*EC: set of characteristics. Eob: set of objects.

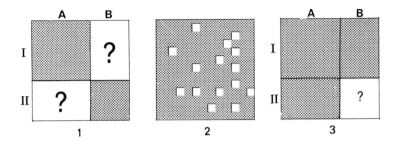

## D.4.3. Description of sampling and interpolation procedure

*The unknown area in a homogeneity schema*

Schema (**8**, p. 238) shows that certain data are unknown. They can be situated in the table in three ways.
(**1**)-*The data are heterogeneous.* The homogeneity schema shows that characteristics I are only known for objects A and that characteristics II are only known for objects B. There is no point in common. The problem is poorly defined. Two solutions: consider them as different problems or look for common elements in A and B or in I and II.
(**2**)-*The unknown data are scattered throughout the cells.* Matrix processing enables us to interpolate the unknown data with a probability that is a function: a) of the proportion of unknowns; b) of the regular or irregular nature of the variation of the components (examples pp. 49, 53).
(**3**)-*The unknown area is homogeneous.* We can see two homogeneous tables: one defined by characteristics I and the other defined by objects A. But the common part enables us to interpolate the missing data with some probability. The unknown area can vary as from (**3** to **5**). The distribution in (**4**) schematizes all the intermediate distributions. Figure (**5**) is the matrix form of the sampling and interpolation procedure. Here are some examples of its use:

- A survey by aerial photography supplies a small number of data items (**I**), but these data apply to the entire surface area (**S**) of a region. On the other hand, agronomists know a large amount of data (**II**), but it only relates to a few points (**P**), analyzed in detail. How can we extend the

## Applications

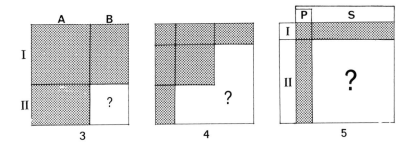

results of the analyses (P) over the entire surface area (S)? In other words, how can we fill in the unknown area of the data table with the greatest probability of accuracy?

- Automatic traffic metering indicates the daily use (S) of the eight access routes (I) to a recreation park. On the other hand, a very onerous manual count gives complete information (II) on the use of each type of recreation but only for several days (P). How do the data (I), indirect but continuous and easily obtained, enable us to predict the use of each recreational activity, that is to extend to every day (S) the results of surveys (P). In short, how can we fill in the unknown area of the table (see p. 51)?

- Statistics provide, for twenty-five million individuals (S) of a country, a small number of characteristics (I). What answers will these individuals give to questions (II), which will arise during a referendum? The samplings carried out for a part (P) of the population attempt to predict these answers. They rely on data I in order to extend the results of the samplings (P) to the twenty-five million individuals (S). They reconstitute the unknown data of the table.

Figure (5) in fact schematizes the main problem in information-processing. What is the information (P) necessary and sufficient for solving problem (S)? The solution is not to provide the total data (S), since these would immediately become information (P) in a higher problem (S') which our imagination would inevitably pose.

## Matrix analysis

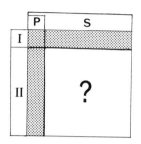

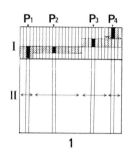

1

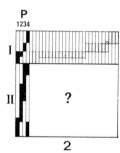

2

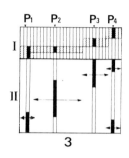

3

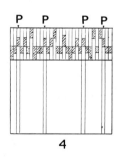

4

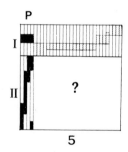

5

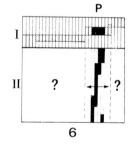

6

Applications 259

*The representativeness of a sample*

Area P corresponds to the elements of the micro-analysis, the significant anecdote, the monograph. Area S corresponds to the macro-analysis, to the generalized level of the study. Obviously, the probability characterizing the correct reconstitution of unknown data, and consequently characterizing the conclusions reached by generalization of the study depends on the representativeness of P in relation to S, across data I. Accordingly, any problem thus posed must start by considering the whole of the table, including the unknown area, in order to focus first on the representativeness of P.

This representativeness has a simple matrix form. Let us suppose that example (1, p. 33) is extended to $n$ communities instead of 16, and that we carry out a detailed investigation II in a reduced number of communities, P1, P2, P3, P4.

If communities P1, P2, P3, P4 are distributed as in (1), the detailed information II can be interpolated over the entire area S with a high probability of accuracy.

But it is obvious that we must first process the data I over the entire area S. We can then choose at least one P per discovered group (1) which ensures its representativeness for characteristics I (stratified sampling).
The Ps are then regrouped and PII is simplified (2). The obtained information is then extended to S on the basis of information I (3). The set of characteristics thus partitions the set of objects (p. 51).

If, on the contrary, we choose P without previously simplifying I (4), there is a possibility that P and S will be simplified as in (5). Interpolation is then very limited (6), and the greatest part of SII remains unknown.

For a number of objects on the order of a thousand, the method of the statistical card, directly applicable to an image-file (p. 88), enables us to make the main differential characteristics appear and to choose P by successive selections based on these characteristics.

D. 4.4. TABLE OF CONTENTS FROM A STATISTICAL YEARBOOK

A statistical yearbook is generally composed of a collection of tables comparing various intersecting components. Any given component thus has a strong chance of appearing in several different tables. This dispersion makes the search for a precise piece of information difficult. And it makes the discovery of comparable series dispersed throughout the book even more difficult.

Can we construct a "table of contents" enabling us to *a)* see if the information being sought does appear in the yearbook, without having to leaf through numerous pages; and *b)* see all the comparisons which the yearbook offers? To achieve this, we must display in minimum size:
1. the list of all the components, and for each one
2. the list of all the components with which it intersects.

This is simply an apportionment table. But when the number of components is very high, the apportionment table exceeds the acceptable dimensions for a table of contents.

An experiment* done with the Algerian Statistical Yearbook for 1972 uses an apportionment table to describe the yearbook **(1)**. Through its horizontal rows of crosses the apportionment displays the main common components. It thus enables us to conceive a "table of contents" **(2-7)** of acceptable dimensions, based on these rows of crosses.

Columns **3, 4,** and **5** are reserved for the most frequently used components. These are generally time, geography and here the distinction between the public and private sectors. Column **(7)** depicts specific or seldom used (e.g. sex) components. Most often these components can only be defined by their categories, whose exact number is indicated in parentheses.
Columns **3, 4, 5,** and **7** define the entries of each table. Column **(6)** defines the enumerated population and **(2)** gives the page number.

This experiment underscores the inefficiency of tables of contents that describe the contents by a general title that merely reflects the enumerated population. The construction of a table of contents for a statistical yearbook is not a secondary problem. A preliminary study in the form of an apportionment table enables us:
1) to check the homogeneity of the yearbook and to perceive the potential weakness of certain headings,
2) to choose a more logical organization for the yearbook than the usual series of typical headings,
3) to give a stricter, more practical definition for each of the tables, based on intersecting variables not just on the enumerated population,
4) to conceive a table of contents that can be classed in different ways.

The publication of several classings for the table of contents would greatly facilitate the search for precise information. But above all, it would enable us to discover the homogeneous statistical sets scattered throughout the various tables, each capable of providing the basis for correlations and multivariate analyses.

*In collaboration with M. FERHAOUI.

# Applications

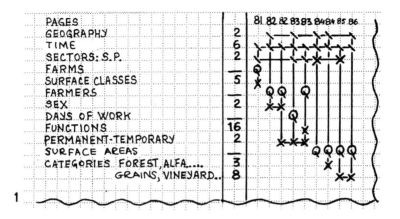

1

| | | | | | | |
|---|---|---|---|---|---|---|
| | o Crop year<br>• Calendar year | W. wilayates (14)  A. areas (4) | | | | |
| | | S. public sector  P. private sector | | | | |
| Pa | 64 66 68 70 | | | | (n.) number  (s.) surface area  (w.) weight | |
| 81 | | | S P | FARMS (n.s.) | / surface classes (8) | |
| 82 | • • | W | | FARMERS | / sex / employers, permanent hired help, temporary (9) | |
| 82 | • • | W | | FARMERS | / sex / farmers, hired help (4) | |
| 83 | | W | S | FARMERS | / permanent, temporary (2) | |
| 83 | o | | S | WORKING DAYS | / permanent, temporary (2) | |
| 83 | o | | S | FARMERS | / perm.temp./ Functions : management, maintenance (15) | |
| 84 | oo | A | | CULTIVATED SURFACE AREA | | |
| 84 | oo | A | | CULTIVATED SURFACE AREA | / forest, alfa, other (3) | |
| 85 | oo | | S P | CULTIVATED SURFACE AREA | / grain, grass, vineyard, orchards, meadows (8) | |
| 86 | oo | W | | CULTIVATED SURFACE AREA | / grass, meadows, orchards, pastures (8) | |
| 87 | | | S P | AGRICULTURAL EQUIPMENT (n.) | / tractors, ploughs, harvesters (14) | |
| 87 | ooooo | | S P | AGRICULTURAL EQUIPMENT | SALE(n) tractors, disk harrows (15) | |
| 88 | | W | S | AGRICULTURAL EQUIPMENT (n.) | / tractors, harvesters, vehicles (5) | |
| 89 | ooooo | | S P | FERTILISER (w.) | / nitrogens, phosphates, compounds (12) | |
| 89 | o | W | | FERTILISER (w) | nitrogens, phosphates, potassics (8) | |
| 90 | ooo | | S P | VEGETAL PRODUCTION (w.) | / grain, vegetables, industrial cult. (45) | |
| 92 | | W | S P | GRAIN (s.w.) | / maize, sorghum, rice, wheat, barley (7) | |
| 94 | oo | | | GRAIN (w.) | / crop, consumption, importation (9)/grain (7) | |
| 95 | o | A | S P | FODDER (s.w.) | / dry artificial, green, meadows, fallow (4) | |
| 96 | o | W | S P | DRY VEGETABLES (s.w.) | / chick peas, beans, vetches, broad beans (6) | |
| 98 | o | W | S P | MARKET GARDENING CULT. (s.w.) | / potatoes, carrots, tomatoes (7) | |
| 100 | o | W | S P | FRUIT TREES (s.) | / olive trees, palm trees, citrus trees (5) | |
| 101 | o | W | S P | CITRUS FRUIT (s.w.) | / oranges, mandarins, clementines (5) | |
| 103 | o | W | S P | OLIVES TREES (s.w.) | / clustered, isolated, bearing, processed (6) | |
| 104 | o | W | S P | PALM TREES (s.n.w.) | / total, bearing /Deglet, Ghars, Degla (3) | |
| 105 | o | W | S P | FIG. TREES | / clustered (s.n.), isolated (n.)/(w.) fresh figs (3) | |
| 106 | o | W | S P | FRUIT (s.w.) | / carob beans, apricots, plums, peaches (9) | |

2   3   4 5   6                                              7

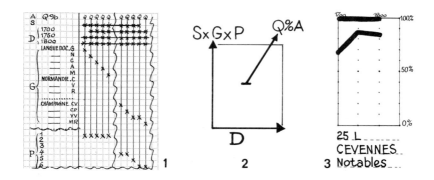

## D. 4.5. MATRIX ANALYSIS OF A SOCIAL HISTORY STUDY

We know the percentage of literates (A)* according to:
- three dates (D): 1700, 1750, 1800,
- sex (S),
- certain regions, cities or villages (G),
- certain professions (P).
What is the effect of these different variables on literacy?

a) - *The data table.* What is the point of comparison, that is the variable which intersects with all the others? In constructing the *apportionment table* **(1)**, we discover as we proceed that any observation, whatever the geographical and professional category concerned, is known across the dates and the sexes. All the information can thus be inscribed in a single table constructed according to the *homogeneity schema* **(2)**.

b) - *The classable graphic construction.* The form of the homogeneity schema leads to an "array of curves." But the difference between the two sexes as well as the "height" of the percentage above zero are very meaningful here. To make them appear, we must transcribe each data item on table **(3)**, with the dates along *x* and the percentages along *y*. The male percentage, always higher, distinguishes the sexes. The region and the profession concerned are written in full on each table. We end up with a *collection of tables*.

c) - *The classing.* We must always attempt to class a collection of tables according to a double-entry system. The classing on the facing page, for example, places geographical items along *x*: different regions of Languedoc (L), Normandy (N), and several regions of Champagne (C). The different professions are placed along *y*: leading citizens **(5)**, merchants **(6)**, craftsmen **(7)**, farmers **(8)**, textile workers **(9)**, agricultural workers **(10)**, and fishermen **(11)**. At the same time it classes regions and professions according to increasing literacy from left to right and from bottom to top.
This classing enables us to focus either on a region, or a profession, or the general tendency. It thus bears an answer to the initial question, while pointing out special cases. Note for example the relative advantages of "Garrigue" for what concerns men but the retrograde nature of its leading citizens, the advantages of workers in relation to vine-growers in Champagne, the advantages of the city in relation to the country, except for Norman "day workers," etc. . .

*Persons who know how to sign their own names are considered here as literate.

## Applications

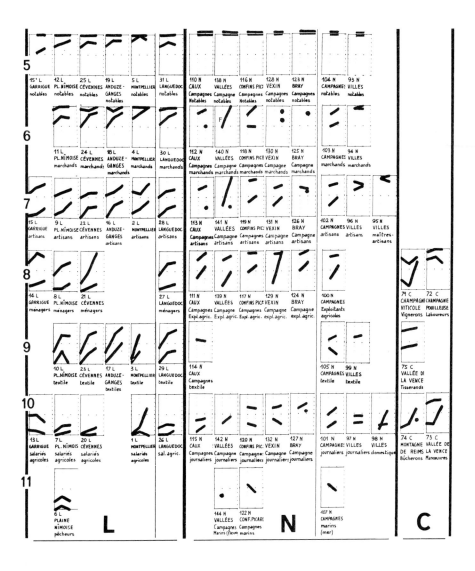

# CONCLUSION

Graphics is a means of communicating with others. This is its best known application. It can also serve to define and resolve problems of information-processing. This application is now going beyond the realm of specialists and becoming widely available, due to a reduction in technical requirements and to semiological simplifications. Moreover, graphics is progressing even farther by giving a visible form to research and methodology.

*What is "research"?*
Research amounts to simple, practical questions for which an answer must be found. Take for example an audio-visual training course for one hundred teachers, each obviously very different from the other. The training will be all the more fruitful if it can give each individual teacher what is most useful to him or her.
On what basis should we divide the participants? What should we teach each group? What equipment should be explained? These are the basic questions.

*What is "scientific research"?*
This is research which reduces the *a priori* by justifying the answers, research which excludes the instinctive answer, which would avoid, in the previous example, dividing the teachers according to primary or secondary level or according to subjects taught. *It involves becoming "informed"* as to what differentiates the individuals. This means a questionnaire, answers, and the implementation of *information-processing* with

no other purpose than to make the way that the answers group individuals and characteristics appear.

*It then involves "interpreting,"* that is giving a meaning to the groups that the processing made appear. This means discovering for example, that a first "system of characteristics" divides the participants into "born teachers" and others; and that within each group a second system contrasts those responsible for multiple classes (classes with several teaching levels) with those who have only a single grade per class. Since these two systems account for nearly all the teachers, we can usefully define four groups of teachers and thus discover what should be done for each. These are the answers to the basic questions. To what extent are these answers open to criticism?

*What is "scientific criticism"?*
Two-fold:
- *Criticism of the questionnaire.* Were the questions correct? Were the people to whom they were asked representative? *These criticisms can only be objective after processing has taken place.* Consequently, thorough research is accomplished in several stages, but each stage follows the general procedure: questions, information, groupings, interpretation, answers.
- *Criticism of the level of reduction.* In the present example, we retain two systems which define four groups of individuals. However, some might say that this is not accurate, that there are subtle distinctions, that there are many more than four groups! Answer: to be completely accurate there are 100 different individuals. Does that mean that we should define a hundred different "groups"? The level of reduction is determined either by the practical problem: how many groups of teachers can we effectively train separately? Or, in an academic study, by the problem of communication: how many notions can the reader retain?

*A methodology for research*
When the principal investigator has realized—
- that the elements of research—questions and information—can always be inscribed in one double-entry table;
- that this table is permutable;
- that these permutations supply the elements of the sought-after answers;
- that these answers are none other than the interpretation of the groups to which the data reduce the $x$ and $y$ entries of the table;
- that this reduction is the goal of mathematical or graphical data processing—

*Conclusion* **267**

then the investigator has a much better idea of what he must do. Relieved of procedural worry, he can devote his entire imagination to what he alone can offer: useful questions, information suitable for processing, and a great amount of extrinsic information to be introduced at the stages of choice and interpretation.

We then realize that the hardest work takes place before processing: "a well defined problem is already half solved." But to define a problem well, we must understand what is behind the notion of "data processing." Graphics, armed with matrix analysis and visual permutation, provides an answer easy to understand, use, or transpose mathematically.

We can also realize that the real chapters of any study are obviously:
- the exposition of questions and data; that is, the introduction;
- the justification of the choice of processing method, a choice depending on the questions and the data but also on the means and time available;
- the stages of processing and interpretation;
- the answer to the questions; that is, the conclusion.

Finally, we come to realize:
- that the ultimate exposition of the problem and the results, that is the exposition of questions and answers, can be summarized in two pages, to the great advantage of the dissemination of scientific information;
- that the justifications, the manipulations, and the interpretation of the groups and their relationships can generally be displayed in fewer than 100 pages;
- that a longer exposition is only to be imagined by taking into account more anecdotal levels of interpretation, which has only two justifications: providing significant anecdotes useful in communication and highlighting exceptions likely to raise new questions and suggest new research.

But to become intimately aware of all this it is necessary to relearn how to "SEE."
That is perhaps the essential property of Graphics.

# PRINCIPAL EXAMPLES CITED

| | |
|---|---:|
| *Hotel management,* Image-file 12 x 20 | 1 |
| *Employment trends in physics in the USA (1971).* Matrix 19 x 13 | 40 |
| *Prices in 31 major world cities in 1970.* Matrix 129 x 31 | 42 |
| *Siberian hydro-meteorology.* Interpolation matrix 100 x 52 | 48 |
| *Recreational park usage.* Interpolation matrix 54 x 32 | 51 |
| *Mendeleev's classification.* Pedagogical matrix 83 x 27 | 52 |
| *The ionic capital.* Matrix 82 x 78 | 54 |
| *Classing of ancient Chinese vases.* Scalogram 17 x 16 | 58 |
| *Trade of the "Comecon" countries.* Weighted matrix 7 x 4 | 60 |
| *The European electrical industry.* Weighted matrix 7 x 4 | 65 |
| *Statistical data on leisure.* Weighted matrix 23 x 25 | 68 |
| *Crop cycles in an african region.* Image-file 13 x 17 | 72 |
| *Study of animal behavior.* Image-file 96 x 12 | 75 |
| *The size of farming operations in Provence.* Image-file 1000 x 9 | 80 |
| *Regional population pyramids in France.* Image-file 22 x 8 | 84 |
| *Factory workers.* Matrix-file 250 x 12 | 86 |
| *A Survey into audio-visual methods in teaching.* Matrix-file 200 x 31 | 88 |
| *Rates of exchange.* Array 216 x 23 | 94 |
| *Fertility trends in Europe.* Superimposed arithmetic curves | 95 |
| *Harbor traffic at Paranagua in the 19th century.* Array 131 x 86 | 96 |
| *Demographic movements in the 17th century.* Collection of tables | 124 |
| *Portraits of Eskimo hunters.* Collection of tables | 124 |
| *Folk songs.* Collection of tables | 126 |
| *Rural exodus.* Networks | 132 |
| *Financial exchange in a market economy.* Network | 132 |
| *Should we retain the "tree" form?* Network | 132 |
| *French constitutions.* Networks | 137 |
| *Land prices in eastern France.* Maps | 146 |
| *Labor force in France.* Maps | 152 |
| *Crop cycles in an African region.* Maps | 156 |
| *Raw materials for the US chemical industry.* Maps | 158 |
| *Utilization of a collection of maps* | 160 |
| *Doctors, migrations, farmers.* Maps | 162 |
| *Ecological planning in Toulon.* Maps | 164 |
| *Typology of 100 communes in Ardennes.* Maps | 166 |
| *Urbanization of a village.* Map | 168 |
| *Evolution of physics chairs between 1700 and 1761.* Maps | 169 |
| *Ethnic maps* | 170 |
| *Map of tree cover in an African region* | 172 |
| *Rainfall and forests in the Congo.* Map | 174 |
| *Meat production in the Common Market.* Matrix 5 x 5 | 194 |
| *French presidential election in 1965.* Map | 210 |
| *Rural co-operatives in Spain.* Map | 224 |
| *Table of contents from a statistical yearbook* | 260 |
| *Matrix analysis of a social history study on literacy* | 262 |

# INDEX


Anaglyphs *(Anaglyphes)*, 151
Apportionment table *(Tableau de ventilation)*, 19, 235..., 260
Area, implantation by *(Implantation zonale)*, 149, 174, 188, 207, 215, 223
Array of curves *(Eventail de courbes)*, 90
Associativity *(Associativite)*, 213, 215, 231
Automation, computor *(Automatisme, ordinateur)*, 2, 7, 9, 21, 35, 47, 65, 161, 211, 249, 251

Base map *(Fond de carte)*, 141
Bounds of a series *(Bornes d'une série)*, 201

Cartography and data processing *(Cartographie de traitement de l'information)*, 155, 161
Cartography, "mark" *(Cartographie de repérage)*, 155, 168..., 225...
Cartography, thematic or polythematic *(Cartographie thématique, polythématique)*, 140
Cartography, topographical *(Cartographie topographique)*, 141, 143
Characteristic *(Caractère)*, 19, 25, 27, 101, 140, 203, 251, 255
Chart-map *(Cartogramme)*, 155
Circle, diagram in *(Cercle, diagramme en)*, 14, 105, 111, 117, 153, 155, 193
Circles, proportional *(Cercles proportionnels)*, 199, 205...
Class *(Classe)*, 19, 105, 107, 109, 149, 203
Collection of curves *(Collection de courbes)*, 90...
Collection of maps *(Collection de cartes)*, 115, 155..., 193
Collection of tables *(Collection de tableaux)*, 124..., 193, 263
Color *(Couleur)*, 28, 163, 186, 217..., 231
Communication, graphic *(Communication, graphique de)*, 11, 16, 22, 59, 89, 154, 167
Component *(Composante)*, 19, 184, 186, 235
Comprehensivity *(Exhaustivité)*, 16, 22, 154, 157
Computor *(Ordinateur)* (see Automation)
Concentration curve *(Concentration, courbe de)*, 100, 109
Construction, circular *(Construction circulaire)*, 14, 105, 111, 117, 153, 155, 193
Construction, circular networks *(Réseaux)*, 129
Construction, matrix *(Construction matricielle)*, 15, 24, 28, 101, 192, 195

Construction, triangular *(Construction triangulaire)*, 113, 121, 124
Construction, useless *(Construction inutile)*, 15, 195
Contour curve, Isarithm, Isoline *(Courbe de niveau, d'égalité, isolignes)*, 143, 151
Copying *(Reproduction, reprographie)*, 35, 36, 145, 215
Criticism *(Critique)*, 264
Cumulative curve *(Courbe cumulative)*, 105

Data aggregation *(Agrégation des données)*, 19, 253, 254
Data analysis *(Analyse des données)*, 19, 233...
Diagram *(Diagramme)*, 15, 18, 33, 192, 195
Difference/Resemblance *(Différence/Ressemblance)*, 176, 221, 225
Dissociativity *(Dissociativité)*, 197, 231
Distribution diagram *(Distribution, diagramme de)*, 25, 28, 101, 103, 115, 200
Domino equipment *(Matériel domino)*, 35, 43, 49, 55, 75, 167

Extremes, tails *(Extremum)*, 201

Factor analysis *(Analyse factorielle)*, 21, 47, 251
Flow chart *(Organigramme)*, 136

Generalization, cartographic *(Généralisation cartographique)*, 145
Graphic processing *(Traitement graphique)*, 16, 18, 20, 22, 31..., 251
Graphs *(Graphes, mathématique des)*, 129
Grid of points *(Grille de points)*, 207
Groupings *(Groupements)*, 7, 11, 180, 230, 265

Homogeneity schema *(Homogénéité, schéma d')*, 17, 163, 234, 241, 257
Hundredths *(Centièmes)*, 190
Hypothesis *(Hypothèse)*, 12, 19, 246

Identification: external, internal *(Identification: externe, interne)*, 177, 225
Image *(Image)*, 21, 180, 185, 213
Image-file *(Fichier-image)*, 28, 71...
Implantation: by point, line, area *(Implantation: ponctuelle, linéaire, zonale)*, 149, 171, 188, 197, 207, 223, 225
Imposition *(Imposition)*, 129, 192
Index *(Indice)*, 190
Indicator *(Indicateur)*, 19
Information: comprehensive, simplified *(Information: exhaustive, simplifiée)*, 16, 22, 154, 157
Information: elementary, overall *(Information: élémentaire, d'ensemble)*, 11, 180, 213, 264
Information: intrinsic, extrinsic *(Information: interne, externe)*, 9, 20, 139, 265
Information, useful *(Information utile)*, 11, 15, 195
Interpolation *(Interpolation)*, 18, 49, 51, 239, 256...
Interpretation *(Interprétation)*, 9, 16, 20, 23, 41, 47, 57, 66, 77, 85, 185, 252, 264
Intersection of variables *(Croisement de variables)*, 236...
Inventory of the components *(Inventaire des composantes)*, 19, 235
Isarithm, isopleth *(Isoligne, isopleth)*, 151

Laws of Graphics *(Lois de la graphique)*, 183

Levels of information *(Niveaux de l'information)*, 11, 180, 213, 264
Levels of questions *(Niveaux des questions)*, 11, 180, 213
Levels of reading *(Niveaux de lecture)*, 11, 180, 213
Levels of reduction *(Niveaux de réduction)*, 267
Line, implantation by *(Implantation linéaire)*, 149, 188, 207, 215, 223
Linear smoothing *(Ajustement linéaire)*, 95, 115, 121

Maps *(Cartes)*, 29, 139
Maps, "grid" or "mesh" *(Cartes par maille)*, 145, 165 (5-6)
Maps for reading, for seeing *(Cartes à lire, à voir)*, 147
Maps, synthetic *(Cartes de synthèse)*, 163
"Mark Map" *(Carte de repérage)*, 155, 226

Mathematics *(Mathématique)*, 20, 129, 178, 233, 265
Matrix, matrix constructions *(Matrice, constructions matricelles)*, 15, 24, 28, 101, 192, 195
Matrix analysis *(Analyse matricelle)*, 3, 16, 184, 233...
Matrix-file *(Fichier-matrice)*, 28, 86...
Matrix, initial *(Matrice zéro)*, 36, 43, 48, 54
Matrix, interpretation *(Matrice d'interprétation)*, 57, 89, 167, 185
Matrix, reorderable *(Matrice ordonnable)*, 29, 33..., 167
Matrix, weighted *(Matrice pondérée)*, 29, 61...
Mean *(Moyenne)*, 109, 203, 229
Monosemy *(Monosémie)*, 176, 179
Multivariate analysis *(Traitements multidimensionnels)*, 18, 251

Networks *(Réseaux)*, 28, 129..., 183, 192

Ordinal numbers *(Nombres ordinaux)*, 239
Orientation, variation by *(Orientation, variation d')*, 122 (7), 169, 171, 186, 223
Overall, elementary information *(Information d'ensemble, élémentaire)*, 11, 180, 213, 264
Overall relationships *(Relations d'ensemble)*, 1, 13, 27, 181, 183

Pattern *(Semis)*, 128..., 207
Percentages *(Pourcentages)*, 45, 61, 63, 85, 190, 245, 249
Perception *(Perception)*, 176, 180, 199, 221
Permutation equipment (Matériel de manipulation), 5, 35, 70, 89
Permutations *(Permutations)*, 5, 7, 15, 33, 38..., 125, 159, 187, 195
Perspectives *(Perspectives)*, 151
Pertinency *(Pertinence)*, 12, 19, 152, 157, 199, 202, 234, 245...
Physical object *(Objet matériel)*, 139, 231
Pictography *(Graphisme)*, 176, 225
Plane *(Plan)*, 186, 188
Point, implantation by *(Implantation ponctuelle)*, 149, 171, 174, 188, 207, 215, 223, 225
Polysemy *(Polysémie)*, 176, 179
Population pyramid *(Pyramide des âges)*, 85

Probability, improbability *(Probabilité, improbabilité)*, 48, 63
Profile *(Profil)*, 5, 39, 229, 249
Projections *(Projections en cartographie)*, 141
Property of the visual variables *(Propriété des variables visuelles)*, 177, 186, 213, 231
Proportionality *(Proportionnalité)*, 177, 188, 197

Quantities, quantitative characteristic *(Quantités, caractère quantitatif)*, 90, 147, 181, 189, 197, 205, 213
Questionnaire *(Questionnaire)*, 89, 251, 264

Rates *(Taux)*, 190
Ratios *(Rapports, ratio)*, 190, 199, 205
Reduction of the information *(Réduction de l'information)*, 12, 77, 180, 219, 265
Relief in cartography *(Relief en cartographie)*, 141, 151
Reorderable, ordered component *(Composante ordonnable ≠, ordonnée O)*, 20, 28, 33, 71, 91, 147, 186, 231
Repartition diagram *(Répartition, diagramme de)*, 26, 103..., 200
Resemblance/Difference *(Ressemblance, Différence)*, 176, 221, 225

Sampling *(Sondage)*, 19, 87, 251, 256...
Saturation, color *(Saturation de la couleur)*, 217
Scale, arithmetic *(Echelle arithmétique)*, 91, 95, 117
Scale, common (Echelle commune), 211
Scale, extensive *(Echelle extensive)*, 209
Scale, Gaussian *(Echelle gaussienne)*, 229
Scale, logarithmic *(Echelle logarithmique)*, 91, 93, 117, 119, 121, 128, 229
Scale of steps in z *(Echelle des paliers en z)*, 184, 188, 199..., 201, 205, 208
Scale of sizes *(Echelle des tailles)*, 184, 188, 199, 205...
Scale of values *(Echelle des valeurs)*, 184, 188, 199, 200...
Scale, proper (Echelle propre), 211
Scalogram *(Scalogramme)*, 58
Scatter plot *(Corrélation, diagramme de)*, 24, 28, 101, 107, 113, 119, 192
Schema, homogeneity *(Schéma d'homogénéité)*, 18, 23, 240, 251, 255
Screens *(Trames)*, 215
Selection, visual *(Sélection visuelle)*, 123, 168, 213, 221, 225..., 232
Series of proportional circles *(Gamme des cercles proportionnels)*, 205, 208
Shape, variation by *(Forme, variation de)*, 171, 174, 186, 213, 225
Sign *(Signe)*, 176, 226
Simplification *(Simplification)*, 6, 9, 20, 23, 66, 155
Simplification of base map *(Simplification du fond de carte)*, 145
Simplification of network *(Simplification d'un réseau)*, 129
Size, variation by *(Taille, variation de)*, 147, 186, 197, 205..., 213
Spectrum, color *(Spectre des couleurs)*, 221
Standardization of conventional signs *(Normalisation des signes conventionnels)*, 225
Statistical card *(Fiche statistique)*, 89, 243
Statistical object *(Objet statistique)*, 17, 19, 25, 28, 184, 249
Steps *(Paliers)*, 35, 36, 65, 184, 197, 199..., 215
Stereoscopics *(Stéréoscopies)*, 151
Stereoscopic pairs *(Couples stéréoscopiques)*, 151
Superimposition of images, of tables *(Superposition d'images, de tableaux)*, 101, 109, 123, 140, 152, 163..., 182, 215, 217
Symbolism *(Symbole)*, 228

Synoptic of graphic constructions *(Synoptique des constructions graphiques)*, 28, 101, 233
Systems *(Systèmes)*, 6, 38, 41, 56, 85

Table, apportionment *(Tableau de ventilation)*, 19, 235..., 260
Table, double-entry *(Tableau à double entrée)*, 193
Table of steps in z *(Tableau des paliers en z)*, 36
Table, ordered *(Tableau ordonée)*, 101, 123, 193, (13)
Table, pertinency *(Tableau de pertinence)*, 19, 234..., 245
Terminology *(Terminologie)*, 19, 101
Text, interpreting *(Discours d'interprétation)*, 184
Texture, variation by *(Grain, variation de)*, 83, 149, 174, 215
Text, variation by *(Texture, variation de)*, 83, 149, 174, 215
Thermography *(Thermographies)*, 223
Thousandths *(Millièmes)*, 190
Threshold *(Seuil)*, 181, 227
Time series *(Chronique, chronogramme)*, 91, 115...
Topographies *(Topographies)*, 28, 33, 139, 141, 192
Tree *(Arbre)*, 131
Trichromatic analysis *(Analyse trichromatique)*, 163, 223

Unknown data *(Données inconnues)*, 49, 51, 239, 256

Value, variation by *(Valeur, variation de)*, 149, 151, 165, 174, 186, 197, 213, 219, 231
Variables, visual *(Variables visuelles)*, 177, 186..., 213, 232
Visibility *(Visibilité)*, 199, 211, 229

*CIP-Kurztitelaufnahme der Deutschen Bibliothek*

**Bertin, Jacques:**
Graphics and graphic information-processing/
Jacques Bertin. Transl. by William J. Berg and Paul Scott. -
Berlin; New York: de Gruyter, 1981.
    Einheitssacht.: La graphique et le traitement graphique de
    l'information <engl.>
    ISBN 3-11-008868-1

Copyright © 1981 by Walter de Gruyter & Co. Berlin 30.
All rights reserved, including those of translation into foreign languages. No part of this book may be reproduced in any form - by photoprint, microfilm, or any other means - nor transmitted nor translated into a machine language without written permission from the publisher. **Printed in Germany.**
Typesetting: Mary Ann White, Larchmont, N. Y., USA. Printing: Karl Gerike, Berlin.
Binding: Dieter Mikolai, Berlin. Cover design: Rudolf Hübler, Berlin.